MAWSON IN ANTARCTICA

Also by Joanna Grochowicz

Into the White

Amundsen's Way

Shackleton's Endurance

MAWSON IN ANTARCTICA

TO THE ENDS OF THE EARTH

JOANNA GROCHOWICZ

ALLEN&UNWIN

SYDNEY • MELBOURNE • AUCKLAND • LONDON

First published by Allen & Unwin in 2024

Allen & Unwin
Cammeraygal Country
83 Alexander Street
Crows Nest NSW 2065
Australia
Phone: (61 2) 8425 0100
Email: info@allenandunwin.com
Web: www.allenandunwin.com

Allen & Unwin acknowledges the Traditional Owners of the Country on which we live and work. We pay our respects to all Aboriginal and Torres Strait Islander Elders, past and present.

NATIONAL LIBRARY OF AUSTRALIA

A catalogue record for this book is available from the National Library of Australia

ISBN 978 1 76118 059 0

For teaching resources, explore allenandunwin.com/learn

Cover & text design by Joanna Hunt
Cover and text illustrations by Sarah Lippett
Set in 11/17.5 pt ITC Legacy Serif by Midland Typesetters, Australia
Printed and bound in Australia in April 2024 by the Opus Group

10 9 8 7 6 5 4 3 2 1

FSC www.fsc.org MIX Paper | Supporting responsible forestry FSC® C001695

The paper in this book is FSC® certified. FSC® promotes environmentally responsible, socially beneficial and economically viable management of the world's forests.

www.joannagrochowicz.com

For Ziggy – who ventured into the unknown alone,
and returned triumphant

The Australasian Antarctic Expedition 1911–14

Douglas Mawson (Dux Ipse) – Leader, 29 years old

Edward Bage – Astronomer, 23 years old

Francis Bickerton (Bick) – Engineer, 22 years old

John Close (Terribus) – Odd job man, 40 years old

Percy Correll (Mr Fix-it) – Student, mechanic, 19 years old

Walter Hannam (Bouncing Billy) – Wireless operator, 26 years old

Alfred Hodgeman (Alfie/Hodgie) – Cartographer, artist, 26 years old

John Hunter – Biologist, 23 years old

Frank Hurley – Photographer, 25 years old

Charles Laseron – Taxidermist, 23 years old

Archibald McLean – Chief Medical Officer, 25 years old

Cecil Madigan (Madi) – Meteorologist, 22 years old

Xavier Mertz – Ski instructor, 28 years old

Herbert Murphy – In charge of stores, 31 years old

Belgrave Ninnis – Dog handler, 24 years old

Frank Stillwell – Geologist, 23 years old

Eric Webb – Magnetician, 22 years old

Leslie Whetter (Error) – Surgeon, 22 years old

Sidney Jeffryes – Radio operator, 26 years old

Frank Wild – Leader of the Western Party, 38 years old

Captain Davis (Gloomy) – Captain of the *Aurora*, 26 years old

Sir Douglas Mawson (8th from right) and the members of the Australasian Antarctic Expedition 1911-14, at the main base in Cape Denison, Antarctica.

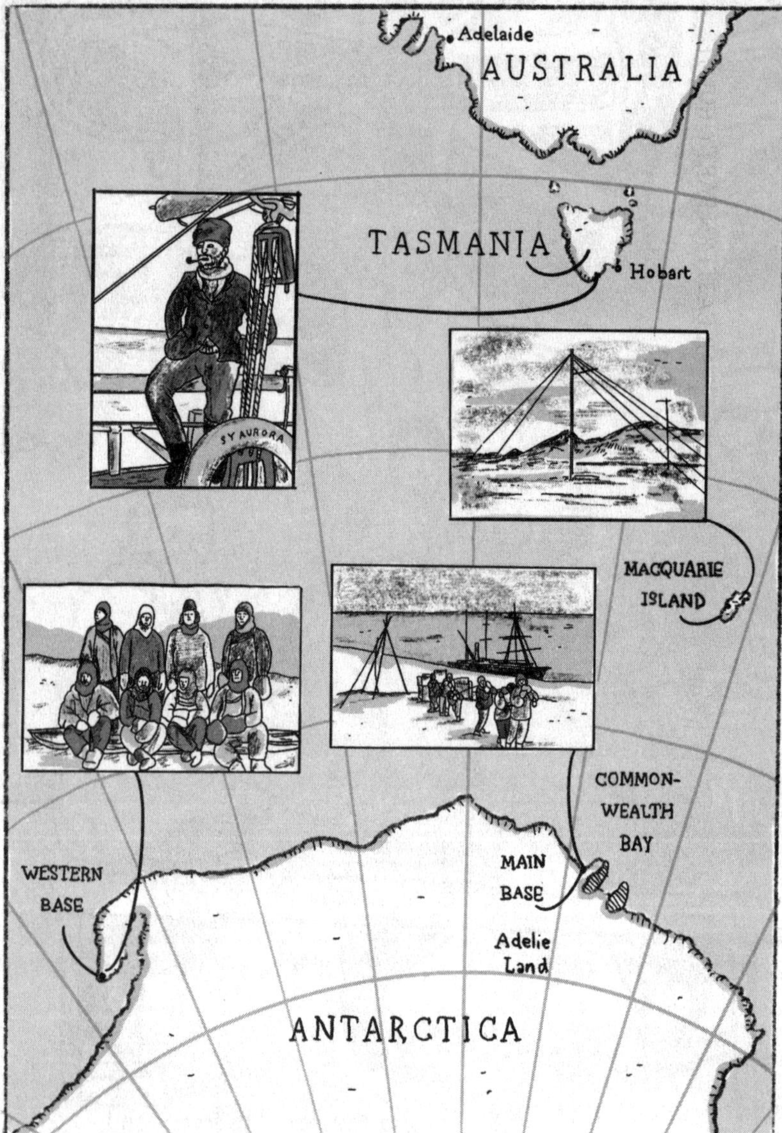
Adelaide
AUSTRALIA
TASMANIA
Hobart
SY AURORA
MACQUARIE ISLAND
COMMON-WEALTH BAY
WESTERN BASE
MAIN BASE
Adelie Land
ANTARCTICA

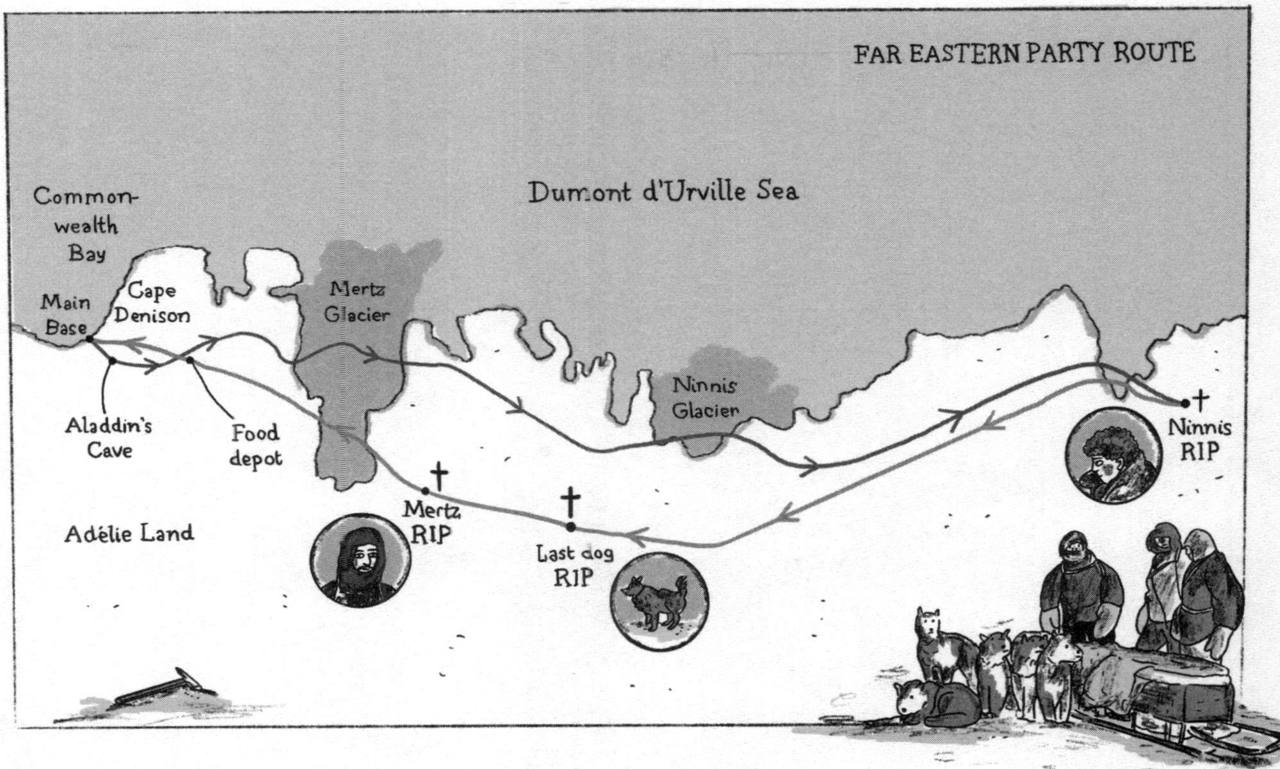
FAR EASTERN PARTY ROUTE
Dumont d'Urville Sea
Common-wealth Bay
Main Base
Cape Denison
Mertz Glacier
Ninnis Glacier
Aladdin's Cave
Food depot
Mertz RIP
Last dog RIP
Ninnis RIP
Adélie Land

PART I

CHAPTER ONE

This ship is no longer on the map. No map exists. This stretch of sea is uncharted. So, too, the wedge of land that appears on the horizon, blank and blinking white. Yet the man standing in the ship's wheelhouse, watching the continent as it shifts in and out of view, recognises what lies beyond the sliding screens of mist and windblown ice particles. Standing here, Douglas Mawson is flooded with memories, very few of them pleasant. Eternal blizzards, blistered feet, foul food is what springs to mind for the young Australian scientist. Antarctica has stirred up a long list of woeful experiences that he fully anticipates having to relive.

Would you like to go to the Antarctic again? people had asked Mawson when he returned to Australia in 1909 after his first polar journey with Ernest Shackleton's *Nimrod* expedition.

Never, was always his swift reply.

And yet here he is, in January 1912, peering through binoculars with something akin to excitement building in his chest. He is a tall man, lean from months of concentration and planning. He has a pleasant face and alert, intelligent eyes, which narrow slightly as he allows his thoughts to slip back in time. After all he had experienced in Antarctica, ordinary life had seemed altogether too easy, dull by comparison. Back in Australia and settled into his university job with nothing more strenuous than the occasional geological field trip, Mawson found himself struck by an idea so urgent and seemingly important that even the comfort afforded by warm food and hot baths could not dislodge it. He will lead his own expedition to explore the uncharted section of the Antarctic coast directly south of Australia.

Mawson turns in excitement to face Captain Davis, a gaunt, red-headed man they all have taken to calling Gloomy. His smile receives nothing but a sober side-eye from the captain. This is a place of tricks and traps. They both know it. What looks like an island turns out to be floating sea ice. And that chain of imposing peaks in the distance will never be climbed. Those unconquerable mountains are nothing but gigantic clouds that will eventually melt before the eyes. To tame such a land, to impose meaning and describe it in scientific terms for the rest of the world, requires a certain boldness of spirit, a single-minded focus, and an appetite for hardship and risk. The captain too has a long list of woes, although they are not the same as Mawson's. He was also one of Shackleton's *Nimrod* men. He is not the least bit excited. Wariness marks the captain's current mood.

'Good to be back,' Mawson says without taking his eyes off the horizon.

Gloomy scoffs at his friend's enthusiasm, although he knows the sea voyage from Hobart has been anything but joyous for Mawson. 'Well, you'll feel good to get off this ship at any rate.'

The rough Southern Ocean crossing has taken a month, with a two-week stop at Macquarie Island to establish a base for the men in charge of the wireless relay station. Heavy seas further south have put men's lives and cargo at risk. Then there are the Greenland huskies. On the days and nights when the waves are so forceful that they have shattered shipping cases, dogs have been found strangled on their chains.

Every task is twice as difficult in a southern swell. Below deck as well as above, waves demand a lurching gait.

On days when the weather runs foul, nothing stays put – certainly not food – neither on the plate, nor in their bellies. Sometimes it flies from the table before even being served, like the joint of beef that bounced across the floor one suppertime and ended up in the third mate's cabin, along with a waterfall of canned peaches and peas and pudding, which came crashing from the wardroom shelves.

'At least you weren't as seasick as last time we came down,' says Gloomy. 'Rolling around vomiting in your sleeping bag like a perfectly useless fool. Then curled up in the lifeboat. As though that would save you!'

Mawson hardly needs reminding. 'I hadn't eaten in days.'

'I fed you up on canned pears. Left you where I found you. That soon sorted you out.'

'And now look where I am. Who would have thought I'd be leading my own expedition.'

'Pity some of your men are not up to much . . .'

Mawson has every confidence in Captain Davis, yet cannot agree with all his opinions, which tend to become gloomier after too little sleep. The young men from Australia and New Zealand who have joined the Australasian Antarctic Expedition – the engineers, scientists, soldiers, doctors, sailors – may not have any experience of frostbite, hardship or isolation, but they all dream of adventure. 'Have to have a little faith in the people I've chosen,' he says.

Gloomy has little faith in anything at present. This feels like the most stressful voyage of his career. His job is to deliver the men and supplies safely ashore. But the closer they get, the more problematic that relatively simple goal becomes. He scowls at the vertical ice cliffs – the continent's insurmountable threshold. With the ship's coal supplies running dangerously low, he is blind to the beauty of the landscape. With each passing hour he is losing hope of ever finding a viable landing site. Ploughing the ship through ice-strewn waters requires both boldness and caution, technical know-how and intuition, which is mentally taxing enough, but the thought that the ship is chewing through his precious reserves to get back to Australia makes Gloomy increasingly irritable. He scowls, barks and swears at any sailor who gets in his way or is too slow following orders. He can be pretty rude.

Gloomy is responsible for getting two survey parties ashore and making sure they're a decent distance apart. There are hundreds of kilometres of coastline to scrutinise in order to find suitable landing sites. And with no maps to guide him, he is navigating blind. This vast region of Antarctica directly south of Australia, known as Adélie Land, was originally scouted in 1840 by a French explorer, but he never made any useful charts or notes, or even set foot ashore. Durmont D'Urville was clearly impatient to leave the place he named after his wife.

'Adélie. Adélie,' Gloomy murmurs to himself, shaking his head. 'Cold and unforgiving. No wonder her husband ran away to sea.'

Mawson thinks of the woman he has left behind in the fondest terms. He doesn't talk about Paquita Delprat with Gloomy or any of the other men. He is intensely private and saves his words of affection for the letters he'll send back aboard the ship. Mawson hopes to marry Paquita on his return. No matter how strenuous and hazardous the field work is or bleak the environment, he will not feel a moment's loneliness with thoughts of his beloved to keep him company. She lends motivation too.

He would like to do her proud, to earn his place in history, to position Australia at the forefront of Antarctic exploration and enable his nation to legitimately make claims on territory. There may be significant mineral wealth to be found here too. Should discoveries be made, Australia should share in it. Yet Mawson's ambitions have nothing to do with winning fame or fortune for himself. Anxious, yet waiting patiently for her fiancé to return, Paquita knows better than anyone: knowledge is the treasure Douglas Mawson seeks.

Another man appears on the bridge. 'Would you take a look at those cliffs!' he crows. The curly-headed photographer called Frank Hurley possesses natural exuberance that is more than a match for Gloomy's dour mood. Mawson is grateful for the comic distraction Hurley often provides. It was apparent from their first meeting aboard a railway carriage in Sydney. Hurley had tracked the explorer down to make him a generous offer.

You don't need to pay me, Hurley had said. *I'll do the job for free.*

Even as the train had pulled slowly out of Central station, and Hurley realised he would have to accompany Mawson all the way down to the Southern Highlands, his banter didn't falter. Mawson had hired him on the spot. Pity Hurley's mother wasn't too pleased. A letter arrived shortly afterwards begging Mawson to reconsider. Mrs Hurley was as fearful as her son appeared fearless. *His health isn't the best. He might not return.* However, he had passed the medical examination. That is not to say that her concerns were irrational. Polar exploration is a dangerous undertaking.

Does Hurley realise what he is getting himself into? Perhaps not. None of these fresh recruits do. And that is just as well. Had they known, nobody would have applied. Those eternal blizzards, badly blistered feet, and that foul food that their leader anticipates experiencing all over again – that is just the tip of the iceberg. Douglas Mawson knows they will come to curse the cold stretch of the landscape as he did. He knows these eager boys will return as men. What he does not yet know is that two will not return at all.

CHAPTER TWO

The three hooded figures huddling on the deck of the ship are doing their best. If only there was more room in the galley to do the dinner prep. *If only potatoes could peel themselves!* It is a miserable chore in such conditions, to have one's hands bare amid snow flurries. With their backs turned to the wind, their numb fingers fumbling with vegetables that freeze before the job is half done, Ninnis, Mertz and Madigan are nonetheless in good spirits. With land now in sight, their longer-than-anticipated sea voyage is almost over.

'How I hate this ship,' says Ninnis as he tosses another potato into the waiting barrel of water, breaking the ice that has formed on the surface. Looking across the stretch of sea to the mysterious landscape on the horizon, he says, 'Finally a chance to step ashore – on land untouched by humans.'

Being here, seeing this. It is a long-held dream for Ninnis. He has spent the last few years serving in the military as a lieutenant. However, since childhood he has longed to follow in the footsteps of his father, whose polar adventures filled him with such wonder while growing up in England. At twenty-four years of age he has had some interesting experiences around the world, but he still possesses such baby-faced features that the others call him 'Cherub'. For years he has tried in vain to get onto an expedition. Having already been passed over by both Shackleton and Scott, Ninnis was overjoyed to finally be accepted by Mawson and would agree to any amount of potato peeling just to be part of this.

Ninnis slips on his wolf-skin mitts to warm up. There's painful metallic throbbing of blood in his fingertips as the circulation returns. He gets a minute or two of relief, then he'll take the mitts off again and get back to work. Despite his present discomfort, he feels giddy with excitement. His anticipation can barely be contained.

'I suppose we should enjoy this torture while we can,' he says to his fellow potato peelers. 'Nothing fresh where we're going.' He cocks his head in the direction of land.

'Apart from the wind,' Madigan quips.

The Australian university graduate is a quick-witted fellow and his wordplay is lost on the third man. Xavier Mertz still needs to work on his English, although the Swiss ski champion frequently raises laughs with his mangled phrasing. Not that he minds when the others find his utterances humorous. Larking about comes naturally. He has kept them entertained by practising his traditional yodelling from atop the mast. Coming from landlocked Switzerland, Mertz is homesick for dry land. Granite summits are what he's used to, not the endless watery peaks and troughs.

Aboard the *Aurora*, Mertz has been helping Ninnis look after the sledge dogs. Over the past six months he has developed new skills in separating fighting dogs, untangling ice-encrusted chains and cleaning up dog filth. Mertz is unable to think of the polite English word for the stuff that gets smeared on his hands and clothing, and carried on the soles of his boots into his cabin; but when he swears in German he certainly finds an appropriate word to describe it.

He and Ninnis are the only members of the expedition who joined the *Aurora* in England. After six months they have become great friends. Physically they could not be more different – Ninnis is lanky and lean, and Mertz is rather short and stocky – yet their names have become forever linked. Interchangeable even. As far as the other men are concerned, it is always *Mertz and Ninnis*, or *Ninnis and Mertz*.

Cecil Madigan can often be found in their company. At twenty-two years old, Madigan is younger than Ninnis, yet he would say wiser. He has postponed a place at Oxford University to join the expedition, and hopes a year off his studies will enhance his academic credentials rather than be a waste of precious time. He hates being unproductive. Even aboard the ship, Madigan has filled his spare hours practising German with Mertz. They write each other notes and correct pronunciation and make fun of each other's childlike errors.

Madigan can think of no better company in which to spend the long dark Antarctic winter than these two fine fellows peeling potatoes. As for the other recruits? Well, not all of them are his cup of tea. There are the ones he finds lazy and loathsome, the ones who are uneducated and coarse, and a few he admits are fine individuals. Sometimes Madigan wonders how they will get along in a hut the size of a sitting room. It's enough to drive a man crazy.

Madigan has given Mawson a Latin nickname. He calls him Dux Ipse – literally, the Leader Himself. It's tongue-in-cheek and not exactly a term of endearment. It conveys Mawson's somewhat aloof nature. None of them quite have a sense of his true character yet. Aboard the *Aurora* he spends most of his time on the bridge with Captain Davis, with both men's attention firmly fixed on the business at hand. Getting the expedition underway and locating a suitable base for their winter quarters is consuming time and all trace of good humour.

Mawson paces back and forth across the bridge, scrutinising the coast that he had hoped would reveal an easily accessible rocky shoreline. The dwindling fuel situation is but one concern. Mawson is fretting that they're tracking well beyond the reach of the wireless relay station on Macquarie Island. There's also the issue of the work he was hoping to complete in the environs of the south magnetic pole. With every passing day, the scope of his scientific program seems to diminish. And with it his ambitions.

When he thinks of the number of letters he has written! Pleading for money, supplies, official approvals, convincing people that there is value for Australia in supporting the expedition. If he scheduled all the meetings he has attended end to end, he would be committed for a month or more. He could quite happily go on exploring for the rest of his life, but organising expeditions he would gladly leave behind.

In that time he has succeeded in obtaining a vessel, manpower, provisions and dogs – all to support an ambitious scientific program. He has travelled to London to discuss his plans with both Scott and Shackleton, negotiated sponsorship deals with businesses prepared to provide food and clothing and equipment at a discount or for free. He has sought out donations from the wealthy and the influential, and the backing of the Australian prime minister, federal and state governments, as well as the learned societies of England and Australia. To top it all off, he has negotiated time off from his university position.

To get this far and not be able to make landfall would be a disaster. He doesn't even want to consider the financial ruin. The expedition's outstanding debts are still mounting. People talk of heroes, of courage and fortitude. But exploring is more than sailing for glory. It's tireless begging, bloody hard work, all followed by much revising of well-laid plans. But Mawson is not alone in this endeavour.

Appearing at the door to the wheelhouse with a swoop of cold air following in his wake, Frank Wild has some good news. 'Can we alter our course? I've seen something.'

Wild is another man Mawson first met on Shackleton's *Nimrod* expedition. There seemed no better choice than Wild to lead the second shore party. Short, wiry and tattooed, Wild has experience, physical courage and pragmatism. Having almost reached the South Pole with Shackleton, he is a man who commands respect. Yet he never shies away from work – dirty or dangerous or just downright unpleasant. All afternoon he has been stationed in the crow's nest – a brutally cold spot for any man to stand, even one as impervious to cold as Wild is known to be.

Gloomy swings *Aurora* in towards the coast for a closer look. With binoculars fixed on the spot Wild has identified, Mawson agrees. Excitement displaces all doubt. Finally some luck. This could be the place.

CHAPTER THREE

The ship is safely moored in an immense bay at least eighty kilometres wide. It has been christened Commonwealth Bay. This is as close to shore as Gloomy dares take the *Aurora* – he can't be sure that the water is deep enough. His reluctance to take risks in unknown waters is even stronger than his burning desire to offload his cargo. A short distance away is a clutch of tiny ice-covered islands, some of them barely more than rocks jutting from the bay. They now bear the name of an expedition sponsor – the man who provided much of their library of Arctic and Antarctic travel books – although it is highly unlikely Campbell Mackellar will ever lay eyes on the Mackellar Islets.

The *Aurora*'s whaleboat moves through the dark water, a first human incursion into nature's wild, perfect splendour. The mood of the men at the oars is boisterous and confident. In the blazing sunshine the landscape appears majestic, breathtaking and welcoming, and they are all overjoyed to be finally making for shore. The continent seems to inflate, as though puffing out its chest. The scale of it is deceptive – as they get closer it expands outwards into an unfathomable distance. Clearly they have misjudged things. Gloomy had assured them that they had barely a kilometre to row ashore – but it now looks at least double that distance.

Mawson and Wild understand that light and the quality of the air alters all perceptions, challenges all pre-conceived ideas about size and scale and dimension of the place. Snatches of sparse conversation pass between them as they point out features of the landform. Mawson has a deep respect for Wild's opinion. He listens, he nods. They seem convinced already – this is the perfect location to establish winter quarters. The others, Madigan, Hurley, Bickerton, Webb and Bage chatter about being the first humans to set foot on this untouched land. They're far more swept away by romantic notions of arrival than the practicalities of actually making a home in such a wild location.

To the east is a wide glacier that tumbles in disarray into the sea. It seems this is where the many icebergs in the bay originate. Further west is a headland that they have named Cape Denison, which offers not only a flat section of land several kilometres across but also a convenient inlet, landlocked on three sides, which makes for a perfectly sheltered landing spot. It seems extraordinarily good luck to find themselves here when the rest of the coastline offers nothing but perpendicular ice cliffs as far as the eye can see.

The place has an odd smell. All remark upon it with wrinkled faces and disparaging comments. Mawson points out such gamey and intense odours emanate from penguins. Sure enough, there's a sizeable rookery. The seal population basking all along the shoreline let off similar rank fumes. It may take some getting used to.

It is not a beach where the whaleboat lands. Rather it's an ice-clad platform standing just proud of the water. It is easy for Mawson to step off and up, effecting the official first landing. He doesn't make any announcements. There is no drama or ceremony. He is simply suddenly on land. An odd muddle of men follows him over, everyone clambering while trying to maintain some decorum, keen as they are to be counted as the second man ever to land. Only Frank Wild waits. He is more interested in tethering the boat. Getting stranded wouldn't be fun. Bad weather always springs from nowhere in Antarctica.

Momentarily taken off balance by the sensation of firm land underfoot, Frank Hurley slips and falls on his backside. 'First to sit on the mainland!' he claims triumphantly.

'Who cares!' someone shouts. Now it scarcely matters. In effect they are all here, all first to walk about on this hitherto unclaimed place – Cape Denison.

Hurley gets his cameras ready, with Bickerton assisting. Wild and Mawson strike off to assess whether the rocky hollow they have spied from the ship could provide a possible building site. The astronomer Bage scoops up a handful of snow. He throws it at the back of Madigan's head. Madigan yelps then kicks snow at the magnetician, Webb, who has barely time to throw his own snowball before Bage slams him with an even bigger one. Their raillery disturbs the deep calm of the landscape and Mawson and Wild pause their conversation to watch, amused at how so-called serious men of science could behave with such childlike abandon.

The snowball war is short-lived; the surface is solid ice with the thinnest layer of snow, all of it too dry, and they're soon out of proper ammunition. Having lost interest, their attention turns to the animals. Penguins, slithering forward on their silvery feathered fronts, approach the scientists. The birds' curious, white-rimmed eyes rove over the humans. How strange to be subject to such scrutiny – shouldn't it be the other way around? The penguins have never seen a human before. Instead of fearful, they are inquisitive. Mawson and Wild only see a food source – for dogs, for men.

*

The men return to the Aurora in the gauzy light of evening. Captain Davis could not be more relieved at the news that they can start to offload immediately; he's less interested in the discoveries they have made – the immense glacier tongue, perhaps the greatest in the world, is of little consequence to him.

'What a feeling – all the tension of the last days is melting away,' thinks Mawson, a grin spreading across his face. 'Well, I can say that I am beginning to feel alive again. And judging by this lot dancing for joy, throwing snow about, so are they.'

The unloading operation will take some time. Work will continue through the Antarctic night. What luck that there's a belligerent sun where a moon should be. The deck is suddenly abuzz with activity and shouting and commands. The sailors get to work unlashing cargo, stacking crates, manhandling dogs out of the way. The new recruits are among it, readying supplies and construction materials – they're as keen to get off ship as Gloomy is to set off. The dogs yelp and bark, perturbed at the sudden winching and hauling and raising of the hatches. The larger motor launch is lifted over the side and placed in the water, loaded with hut timbers, fuel, sledges, harnesses and mutton carcasses.

It is just before midnight when the first relay is sent from the ship. The puttering and spluttering of the motor launch is a jarring sight in the twilight glow, spewing smoke at the effort of towing two such heavily stacked whaleboats behind it.

Frank Wild is keeping a wary eye on the weather. He is concerned that the wind has picked up. He points out to Gloomy how the icebergs have rearranged themselves in the bay, and spindrift now fills the air. Ice crystals twinkle and sparkle – to an unpractised eye it is a delightful effect. But Wild senses something bad. A sleeping giant has woken.

'This is going to get worse,' he says.

Gloomy nods. 'We'll need the anchor to hold well. If we are driven out to sea, those chaps out there will have a fearful time ashore with nowhere to shelter.'

The blizzard sweeps down from a seemingly clear blue sky with astonishing ferocity. The air is suddenly heavy with snow. The waves crest and, turning white, slam against the hull. The wind flings itself about in a chaotic frenzy. Even with its engine at full throttle, the little motor launch has the greatest difficulty forcing its return to the ship. On arrival, Madigan and Hodgeman need urgent help. Their light clothing has set solid with sea spray and frozen them in place. Hodgeman and another man, Hannam, have frostbitten fingers. *What possessed them to dress as though they were headed for a picnic on a sunny, cloudless day?* Wild wonders as he bundles them aboard.

They're the walking wounded, he thinks, marvelling at how none of them recognised the need for appropriate clothing.

What of gloves or mitts? And not one of them with a head covering. No balaclavas or hoods or windproof layer. He's never seen so many exposed ears, lashed and lacerated by the wind. Their lips are blue yet their foreheads are blasted, burned a berry red.

'Get them below. This frozen lot will be good for nothing,' Mawson says to the bosun. 'Keep a watch on them.'

The next watch is roused from sleep before time. Fresh from their bunks, Ninnis and Mertz can see a major calamity is brewing. Quick to lend a hand, they race to hoist the two snow covered whaleboats onto the davits. The frozen ropes score the flesh of their still largely un-weathered, untested hands. They try and coordinate their movements but in the cold their mouths are as clumsy as their fingers. Then there's the bitter struggle against the wind. The confusion of getting everything that was heading ashore re-stowed and secure takes a lot longer than it usually would. And now, to shuttle the dogs under shelter. Howling and pulling against their harnesses and daring to nip at their rescuers, the animals scarcely seem capable of comprehending the need for such sudden rough handling.

Shouts alert Mertz and Ninnis to a crisis overboard; they run to the railing to see the motor launch adrift, with Bickerton and another two men waving frantically. It has broken free from the ship before they could hoist it back on deck. The painter rope trails in the water. The gale carries the launch further and further away. All the while Bickerton attempts to turn over the engine. How will they steer themselves back to the ship against the wind when the rudder has already been removed?

'Haul in the anchor,' Gloomy shouts to the crew. 'We need to retrieve them.'

The shivering, ice-caked deckhands cannot think of anything more unpleasant than to gather in seventy fathoms of frozen anchor cable. Fortunately for them, Bickerton has managed to start the motor launch and navigate it back alongside the ship on an odd meandering course. He's bent double in the wind as he secures the line; the others are ashen faced as they are helped back aboard. More frozen men.

Catching the leader's eye, Frank Wild shouts over the sound of the wind. 'These men! Nobody has any idea, do they?'

'These are the conditions to test a man,' Mawson calls back. It has already occurred to him that he will only know in a year's time who will prove themselves of most value to the expedition.

Wild grips Mawson's arm in a gesture that is partly in friendship, partly for emphasis. 'If you get every one of these fools back to civilisation, it will be a bloody miracle.'

NIMROD

1907 – THE PROF

Field work takes Mawson out across the saltbush plains, far from the dusty settlements, towards the Barrier Ranges. Always alone. He rides a horse, he looks through the lens of his new Zeiss camera, he studies the arid and hostile environment. South Australia is extending the frontiers of his geological knowledge; the accumulation of sediments in the region around Boolcoomatta tells him – glaciers created these landforms.

In a hotel in Broken Hill he unfolds a letter from the Prof, his mentor and former lecturer at the University of Sydney. The Prof writes that he will join Shackleton's forthcoming expedition to Antarctica to further his own studies of glaciation. Professor

Edgeworth David is a man known for his impeccable manners, gentle humour and unfailing support of his former students. He has suggested that perhaps Shackleton would offer Mawson a place on his expedition.

Mawson imagines the thrill of seeing a continental ice cap in real life; it would be like seeing an ice age in action. He would do anything to see great glaciers tumbling into the sea. He would be thrilled to make the round trip on Shackleton's ship – the 300-tonne, square-rigged Nimrod – even if only there and back, working for free.

Mawson finds Shackleton an interesting personality on first meeting the explorer. For his part, Shackleton believes the young geologist would be an asset to the expedition. The job is his if Mawson wants it. The job offer is met with puzzlement, however.

'I'm to be the official physicist for the duration of the expedition,' Mawson says to the Prof.

The Prof beams. 'Excellent. Excellent.'

'But I'm not a physicist.'

'A minor detail,' the Prof waves away the concern. 'Don't worry yourself. Shackleton doesn't care. There will be ample opportunities for geological work. Leave it with me.'

CHAPTER FOUR

Unloading of *Aurora* is resumed two days later, after the worst of the weather front has passed. The blizzard has dispelled all romantic notions of Cape Denison being anything other than a bleak and exposed outpost. Blowing out from the interior, tearing across the water, whipping up waves, the wind gets stronger with each passing day. It is an almost constant harassing presence. In the absence of other options, this is to be their home. There is so much to do to make it habitable.

The heavy cargo is winched from the hold, the lighter boxes handed man to man along a human chain across the deck and into the two waiting whaleboats, which tip from side to side every time another crate zips down the gangplank. Shouts ring out. *Look out!* Heads, legs, boats. To be on the receiving end is dangerous, especially when the cold makes false moves from frozen hands more likely. Still, to be on the move means staying warm at least. Each boat is loaded in twenty-five minutes.

The landing spot looks more like a jetty now. It has acquired an official name: Boat Harbour. A group of men have cut mooring poles into the ice and the gear and building materials are piled high, waiting to be sledged several hundred metres inland to where the building platform has been established.

Nothing can soften the bitter cold. It seeps into their clothes like liquid. Proper winter clothing is very much in style now. In fact, most men have gone for over-the-top interventions, wearing both balaclavas and wind helmets. Nobody would dare to get about with naked ears and a bare head after the fearful wind-lashing delivered by the storm. Similarly, hands are gloved and mitted, although they're all discovering how frustrating it is trying to do anything requiring dexterity. Already a fumbled crate has been lost over the side and now sits in plain view under five feet of icy water. The contents are likely waterlogged and ruined. They won't retrieve it.

The harsh cries of penguins fill the air. The men barely register the sound now. Similarly the penguins have seemingly accepted the odd noises of humans calling to each other and care little about the sudden influx, the strange goings-on down by the water's edge at Boat Harbour or the structure that is slowly taking shape further inland. A few nosy individuals gather to watch the unloading, chattering among themselves for a time before wandering off, nonplussed.

Alfie Hodgeman has been given a puzzle to figure out: how to build their hut. He is a bit of a puzzle himself. He seems to have more jobs than anyone else. A former architect and draughtsman, Alfie has a talent for sketching and map-making, although officially he has been employed as assistant meteorologist. He is yet to register any details about the weather at Cape Denison beyond noting down the basics in his diary: *fine day, winds gusty; temperature 6 degrees at noon, minus 34 during the night.* He hasn't had time. As the only member of the expedition with building expertise, he has found himself suddenly in charge of the construction of the hut.

The trickiest bit arises from the fact that there are two huts that need to be consolidated into one dwelling. The huts were pre-fabricated and erected while still in Australia, but never intended to be joined up as one structure. Furthermore, none of the pieces have been numbered, so nobody quite knows how it all fits together.

There are no kitset instructions, only a vague logic to the walls, floor, windows and cladding. It must be weathertight without any gaps. Given the windy position and the need to maintain structural integrity, the whole thing should be fastened using heavy-duty fittings, but no bolts have been packed. Nails will have to suffice; they'll need a fair number to ensure things hold together. How many of these men have used a hammer before?

Mawson is impatient. He sets a cracking pace. Now that they're ashore, he wants to see progress. Work will continue from seven in the morning until close to midnight. They need somewhere to sleep. Tents are only a temporary fix. And nobody enjoys eating outdoors. Food gets cold very quickly. The bowls of stew that were piping hot when dished up from the camp stove are lukewarm once they hit the tongue.

Hannam has taken charge of the camp kitchen. Bouncing Billy they call him – he's round, jovial and bearded like a teddy bear. He is not the world's best chef but his enthusiasm more than makes up for his lack of experience. Like Alfie, he has been put in charge of a job that is quite unrelated to his official expedition role of radio operator. He's been appointed blacksmith too. Someone needs to repair the tools – in the freezing air the metal drill bits split and shatter with irritating ease.

While he can generally be found bashing away at a forge set up on the rocks, he's getting very good at fixing meals at top speed. It's not really cooking, more opening tins and tipping the contents into a giant pot, stirring it until it looks reassuringly brown and unappetising, then dishing it out. Regardless of what goes in, the same slop comes out. Nobody cares. Physical work, hefting construction materials, manhandling rocks to clear the building platform in the cold has made them greedy, even for gloop.

Bouncing Billy yells, 'Grub's up!'

Forget table manners. At this stage, there is no table. The men balance plates and mugs on packing cases and set upon their food. Hurley jokes that there is little difference between them and the hungry dogs. Just as the dogs seem to block out all other concerns in the world when presented with their chunks of seal meat, so the men fall silent and eat with single-minded intent. All that can be heard is the clatter of cutlery and the odd appreciative sound. They are ravenous.

Bickerton the engineer leans back from his empty plate, satisfied that his supper didn't get cold before it reached his mouth, yet disappointed that it's gone so quickly. There's still cocoa and biscuits, but they too disappear in a matter of seconds. Some choose to round off the meal with a pipe. Bickerton slips his hands into his pockets for the pouch of tobacco, but instead he pulls out a stick of dynamite. It might not be the wisest thing to store explosives in one's trousers, but these are not normal circumstances. Dynamite doesn't behave as it should in sub-zero temperatures. He needs to warm it up if it is to do its job of blasting the bedrock so they can lay the hut's foundations.

'Bick,' says Hurley. 'Don't go lighting your pipe off that thing.'

He laughs. 'Forgot about that.'

The environment is teaching them a valuable lesson: the old established ways of doing things can no longer be trusted. The concrete for the footings isn't setting. The wooden posts that they're using will need to be wedged with bits of timber and weighed down with tonnes of boulders manhandled into place. Of course, nobody can wield a hammer while wearing reindeer mitts, and grasping nails is too painful for bare hands. Occasionally a new discovery saves the day. Madigan has discovered that in the absence of clay, dried penguin poo is the perfect material with which to tamp down the sticks of explosive when placed in the holes they've drilled into the rock.

Even the act of walking takes on a strange new complexity. Underfoot, the wind-polished surface has become as slippery as an ice rink. It is virtually impossible to move about without crampons, especially when the upper body is so harassed by gales. Not even snow can adhere to such a slick horizontal plane, and it flows freely over their feet like a fast-running river. The effect is disorienting. *Am I walking forwards or backwards?* they wonder before losing balance and falling over.

What a place! There are sure to be many weird and wonderful phenomena awaiting them inland. Whenever anyone's gaze is drawn southward, beyond their tiny toe-hold of a camp on the coast, the landscape is utterly unreadable. It looks like an early draft, a place so new that it has yet to develop features or decide on a colour. In truth, this is an ancient land whose wilful and unpredictable character was determined long ago. And it cannot, it mustn't, be trusted.

CHAPTER FIVE

Somehow their first ten days have passed without any major mishaps. During that time the *Aurora* was a reassuring sight, moored a far distance off shore while the last unloading was completed. Seeing the ship disappear over the western horizon, many of the men are struck by the realisation of just how cut off from the world they now find themselves. Of course, Wild's men will be settled further along the coastline, but at more than a thousand kilometres distance, they might as well be on the moon – they won't be calling in for supper any time soon.

It will be a full year until they see Gloomy again. Assuming, of course, the ship returns. Should disaster strike it on the way back to Australia, taking captain and crew to the bottom of the sea, the whereabouts of Mawson and his men will remain forever a mystery.

The ship carries with it a mailbag full of their final letters home. All the men have written farewell letters to friends and family. Mawson himself has written four letters to his fiancée, Paquita. One from Hobart, one from Macquarie Island, one from the Southern Ocean, and finally one from their new home, Cape Denison. He would have easily written four hundred, but his time has been so desperately short, with the need for close supervision of the new recruits.

All eighteen men are squeezed into four tents - with only canvas and bamboo tent poles to protect them against the Antarctic elements. The rest of the camp is a jumbled mess, with crates containing two years' worth of food, 23 tonnes of coal, scientific equipment, a workshop's worth of heavy tools, household items like plates and cups and cutlery, candlesticks and camp chairs, their sledges, winter clothing. A short way off lie the three radio masts that still need to be shifted into place. And the aeroplane.

The Vickers plane is Bickerton's responsibility. He's not the pilot. Originally Mawson had hoped to fly it, but the plane no longer has any wings, after suffering a dreadful crash in Adelaide. Poor Frank Wild was a passenger. Thankfully nobody lost their lives; the wings were the only casualty. While it will no longer fly, the wingless plane, which has been christened *Grasshopper*, will perform another function. With a few well thought-out adjustments, the *Grasshopper* will be used to haul sledges as a strange sort of hybrid air tractor. Bickerton will have to figure it all out, though Hurley will lend a hand. He too has excellent mechanical sense and a knack for inventing, repurposing and rejigging. Except when there are critical pieces missing.

'I just can't find these fittings for the stove.'

'They must be somewhere, surely,' Bick says.

Crates are jimmied open, rifled through. A crowd of men go looking. Everyone wants the stove to work – for cooking, for warmth. Boxes are upended onto the snow as the desperate hunt goes on.

'Just our luck. I bet you the missing pieces are in the crate we lost in Boat Harbour. Do you think?' Hurley's face assumes a look of pure dread.

Word goes out. Nobody wants to jump in the water to retrieve it. They have already attempted to haul it up from where it sits on the sea bottom, taunting them amid the thick kelp. It's sitting under more than a metre and a half of icy water. There's only one way to do it.

'I'll retrieve the wretched thing.' Mawson is the only volunteer. They may be calling him Dux Ipse – the Leader Himself – behind his back, but Mawson does his fair share of tough jobs. Already he has won the respect of his men by never asking anyone to do anything that he was not prepared to do himself. He has one request, however. 'I'll do it when the tide is low. No point more of my body having to suffer than is necessary.'

The next day, before breakfast, an audience gathers to watch his skinny dip. Sliding into the water, he yelps loudly. The dogs get going themselves, convinced that something terrible must be afoot. The men watching his ordeal cheer him on, grateful it's not them swimming about stark naked. After a bit of awkward fumbling and trying to avoid putting his head under, Mawson finally gives in. Three times he dives. Three times his head bursts free from the water, taking dramatic gasps as though the shock is almost too much for words. On the fourth attempt he succeeds in bringing it to the surface. The troublesome crate is finally on dry land. Then Mawson himself is hauled back onto the ice by a flurry of proffered hands. He leaps back into his clothing in what seems like three seconds flat.

'That must be a world record in dressing,' Hurley observes, rubbing his leader's arms frantically as others slap him on the back. Someone throws a blanket about his shoulders.

Bickerton wrenches open the lid of the waterlogged crate. His face drops. 'Well, we've located the strawberry jam.' He says dejectedly. 'One hundred and forty cans of it.'

Hurley roars with laughter. Bick groans. They need a working stove. Their survival depends on it.

'Let's not give up,' Mawson says sternly. 'Those parts are somewhere.'

Some may think of Mawson as an academic, an intellectual, but his physical courage cannot be doubted. His utter and selfless dedication to achieving his objectives is something to behold.

CHAPTER SIX

The foundations and the outer framing of the hut are finished. When the sharply raked pyramid-shaped roof goes on there's whisky and hot chocolate to celebrate. Soon enough, the rafters, the cladding, and the flooring are finished, and the interior panelling can be installed. Madigan and Ninnis volunteer to squeeze themselves into the space under the eaves and attempt to nail in the ceiling panels while lying on their backs.

The structure looks surprisingly solid for something erected by men with no building experience. There is still some anxiety, however. What will the wind make of their handiwork? Despite the blazing sun, the intermittent gales are felt most keenly – especially on the bare hands and faces of the men working on the roof. Yet the view from up there is marvellous. Their knowledge of the surrounding area extends only as far as the naked eye. They have all clambered up top to survey the rocky ridges that form the natural boundary of Cape Denison, the glacial lake that extends out from beside the hut, the steep ice slopes rising steadily to meet the colourless plateau that lies due south. It seems to be the only road for all further explorations of the landscape. Eventually they'll all set forth in that direction.

A few hammered fingers, scrapes, strained muscles and splinters are the only injuries, although they are all displayed with something approaching pride. Up until now they have been exceedingly lucky. There have been no accidents with deadly consequences – only comical incidences with painful bouts of belly laughter.

Sitting astride the roof, Bick had a moment of high drama when Madigan accidentally drove a nail into his backside while working underneath him in the hut, hammering in the ceiling panels. Taking fright at Bick's cry of pain, poor Mertz tumbled backwards off the roof, pulling the entire chimney stack off with him and howling all the way down. Neither in English nor in German, but in some universal language of distress, Mertz's cries were immediately understood by the dogs, who took up the chorus with enthusiasm. Thankfully, the impact of his fall was softened when he landed in a snowdrift; he was only winded. Meanwhile, Bick's wounds need tending by Dr McLean, who everyone now calls Dad.

'Let Dad take a look at your punctured buttock,' Dr McLean says calmly.

Bick winces. Despite there being no ill-will on the part of Bick, Madigan hovers with a look of awkward sympathy. It seems anyone working alongside Madigan is cursed. Later on there are more cries of pain when Hunter's fingers are crushed between a rock and a heavy crate he was carrying with Madigan.

'Dad! Medical kit!'

The fingers are a mess. Sewing them up, Dad cautions Hunter to take it easy. The biologist will have to use his left hand to write, not that he has any time for doing science. He will be on light duties for more than a week. The good news is, the heavy crate contained the missing bits to the stove. It can now be installed by Hurley and Bick, while a team attempts to construct two tiers of bunks and shelving to line the interior of the hut. Most of their supplies will be stored in the covered verandah that extends around the exterior of the building. Once the snow accumulates around the hut, this area will be completely enclosed.

It's going to be a tight squeeze for eighteen men. Only Mawson will have a room of his own, although it is more like a cubby hole, with a desk and a bed, separated by a short partition wall. Everyone else will bunk down in the main room. It is only to be expected that the leader of the expedition should enjoy a measure of privacy. It also gives the men the impression that the boss is not always breathing down their necks, even if it is merely an illusion. Mawson will be able to hear every word spoken by the men sharing this small hut. If anyone wants real privacy, they might only find it in their diaries. Then again, there is an entire continent offering privacy for anyone who cares to step a few metres from the hut.

Mertz won't chance it on the roof again. He is far happier on firm ground wrangling the dogs with Ninnis, although the main job now seems to be stockpiling food for them. Hauling sledges after being inactive for so long at sea, the dogs have ravenous appetites. Only huge helpings of seal meat will satisfy their greedy howls. Ninnis and Mertz are less greedy for the hateful task of butchering. It is cold, messy and smelly. With such limited options for washing, the unusual odour of blubber tends to cling to clothing and skin.

The seal population has expanded; its sprawl of tawny, steaming bodies coats the coastline like an oil spill. What luck to have such an abundant food source right on their doorstep. There is even greater excitement in the afternoon when a gigantic elephant seal launches its quivering mass onto the shore. It is a monster of a thing; coughing and belching as though announcing its arrival. Too late, it realises an ambush is what lies in store. For the men, an elephant seal - a rare beast this far south - represents fantastic luck. It is an opportunity to secure enough meat for the dogs for an entire month.

The bonds of friendship are growing stronger every day between four men in particular - Ninnis, Mertz, Madigan and Bickerton. The four of them set to on the carcass, although none of them really has a clue about technique.

Somebody's knife slices inadvertently into the terrifically offensive guts, and dreadful smelly gas engulfs them. Mertz turns away in disgust, put off completely. Ninnis dry retches. Great pockets of putrescence rise into the air from the stagnant waste of the creature's stomach, which seems to contain great quantities of decomposing fish in various stages of fermentation. Oddly, it seems to be the dogs' favourite part of the animal. They throw themselves upon the heaping mess in a frenzy.

'Don't feel you need to leave any for me, lads,' says Hurley dryly. He's been photographing penguins. The Adélie penguins are more stroppy than the king or gentoo penguins the men saw in their thousands on Macquarie Island. The Adélies are not afraid of giving a mighty slap to any man who gets too close. There are still chicks about at various stages of moulting. Threadbare and long suffering, they wait in the rookeries for their tufts of fluffy baby feathers to be torn away by the winds.

Lying in his reindeer sleeping bag that night, his last in the tent, Ninnis records the day's activities. It takes energy and enthusiasm to note everything down when he is physically exhausted. Yet everything is new. Everything seems noteworthy – especially the fact that his ink has frozen and his fountain pen needs to be warmed up repeatedly in order to write.

'Mertz,' Ninnis pauses over his diary, 'what's the date?'

'I don't know,' comes the Swiss man's drowsy response. 'I know one thing. When I am not working I am sleeping.'

Ninnis supposes it does not matter greatly. The sun never sets. One day melts into the next. Tomorrow there will be more work, more laughter, more exhaustion and definitely more wind. Always more wind. It screams down from the interior and only seems to gather speed as it is funnelled between the rocky ridges of Cape Denison. Sometimes it stays awhile, prowling menacingly around their hut before fleeing north, its rage spilling seaward. Suddenly the water is churned from blue to white, and waterspouts dot the surface like wobbly towers. Icebergs are chased far from the coast as though their mass and weight pose no obstacle at all. Icebergs are no match for the wind. And neither are the men.

NIMROD

1908 – EREBUS

It's the Prof's idea. He's not a fit man, nor particularly young anymore, but enthusiasm has taken hold. To be the first to scale an Antarctic volcano would be a marvellous endeavour, would it not? He seeks permission from Shackleton, and insists that Mawson is his man.

'Bring your camera.'

'How high is Mount Erebus?' Mawson asks amid the geological specimens, instruments, notebooks and packing cases that line their sleeping quarters.

The Prof is lying on his back in bed. He squeezes one eye shut and gives his best estimate. 'Three and a half thousand metres would be my guess.'

'Do you have mountaineering skills?' asks Mawson.

'No!' The Prof scowls, sounding almost offended that such a trifling detail should come into it. 'And neither do you.'

Another man, Mackay, will join their summit party. And three others – Adams, Marshall, and Brocklehurst – will help haul the sledge they have packed with ten days' provisions.

'Erebus is erupting!' shouts Mawson a day before departure.

True enough, an immense mushroom cloud rises a few thousand metres into the air.

'What luck!' says the Prof. Booted, suited, with ice axes in hand and crampons improvised from leather and nails strapped on as best they can, they set off.

Three days in and the going gets tough. It's minus 30 degrees Celsius, the snow is deep, the gradient is increasing. They abandon their tent and sledge, and instead cut footholds in the steep ground. Mackay faints from exhaustion; Brocklehurst's toes turn black; they're caught in a blizzard. For a time they proceed on their hands and knees. Breathless, they give up on conversation. But not on the climb.

It takes them six days, but they reach the top amid the sulphur spewing forth and the aggressive boom and hiss of the volcano. Mawson and the Prof inch around the crater. The Prof points to the feldspar crystals, Mawson takes photos. There are scientific readings, geological samples to gather. Then the Prof suggests the unthinkable: a glissade down the icy slopes all the way to the bottom.

'Ice axe for brake, what do you think?' The Prof offers brightly.

They descend at speed, slipping and sliding hundreds of metres, until they arrive exhausted at their first depot. Back at the hut the following day, the heroes are feted.

And just like that, the first ascent of Mount Erebus is under their belt. Mawson feels great. He suddenly feels physically capable. He wonders what other firsts lie within their reach. Already there is talk about bagging the capricious, ever-roving magnetic pole. But it's geologising that he craves, and the Dry Valleys beckon.

CHAPTER SEVEN

Mawson sits at his desk and admires the elegant menu that Alfie has designed to mark the occasion of their first dinner in the new hut. Across the top of the page, the Australian Antarctic Expedition initials, AAE, take pride of place on an unfurling banner proclaiming *Advance Australia.* Below it, the evening's four-course meal appears. Alfie has certainly had fun concocting fancy-sounding names for what amounts to a pretty basic meal of penguin, peas and potatoes. The so-called seasonal vegetables are a slight exaggeration – then again, tinned vegetables are always in season in Antarctica.

Roast Fillet of Dux a l'Adélie
Legumes in season
Iceberg blancmange
Nunatuk Jelly with blizzard cream

However ordinary the food, he is sure everything will taste marvellous when they are seated together at the long table that Dr McLean and Stillwell have constructed. It now occupies the only spare space in their living quarters. And music! There is sure to be lots of singing when they crank up the gramophone for the first time.

Mawson wonders if Captain Scott has been successful in his quest, or if Amundsen has beaten him to the southernmost place on Earth. Should neither Scott nor Amundsen find success at the pole, Shackleton will surely make another attempt next year.

If Mawson had accepted Scott's invitation to join the *Terra Nova* expedition, he may have been among the first men in history to reach the South Pole. There was a time when Mawson had considered it. Had Edward Wilson not already been appointed chief scientist, Mawson would have liked the top scientific job for himself. It is not necessarily the stuff of newspaper headlines. The public generally prefer to read about world records rather than scientific discoveries.

But for Mawson the key goal is charting this section of the coast - putting it on the map and in so doing, establishing a claim for Australia. It's a huge area to explore; the sooner the various sledging parties can set out in various directions in spring, the greater the likelihood of them contributing meaningfully to knowledge of the continent.

There had been a time when Mawson thought Shackleton might lead the Australian Antarctic Expedition. He was an early enthusiast, promising to help with funding and putting Mawson in touch with sponsors in London. Ultimately, Shackleton's enthusiasm for taking part had waned. His view was that, assuming the pole was reached, the only worthwhile polar endeavour would be the crossing of the continent - another world first.

One world first that Mawson is passionately interested in, is establishing the first communications link with Antarctica. He is impatient for progress. Once contact with the relay station at Macquarie Island is achieved, they will be able to send and receive news. They may even learn if Scott has been successful.

So far they haven't been able to erect their radio masts. The wind has scuppered all efforts. It is proving a wild and constant presence at Cape Denison. Blowing even on brilliant blue days, the wind robs the sun of any warmth and is beginning to rob the men of any eagerness for outdoor tasks. While it can be very pleasant to bask like seals in the sun's rays in a sheltered spot, to stand in the full force of the wind is to offer up one's face for a thorough blasting with fine particles of ice.

Skiing has occupied more of their time of late, and the wind can be of assistance if it is blowing in the right direction. Mawson is pleased to see Mertz putting all the beginners through their paces. Any practice they can get while there is still daylight will be an advantage when it comes to setting out on their survey journeys in the springtime. Falls are inevitable with so many young men trying to outdo each other. Most are slapstick, some are spectacular. Hurley captures many of them on film. He gets the men to ski off an icy bank just so they will crash – he's already thinking of how best to entertain his audience back in Australia.

Ninnis bruised his tailbone so badly it feels like a hammer hit it. Dr Whetter, the man they have taken to calling Error, had a major tumble coming at great speed down the steep slope that leads out of Cape Denison and onto the polar plateau. He smashed his shoulder in his blaze of glory and now has nothing but grief. Dad is called. His verdict: Error's injury is bad enough that the others will need to take over his duties for a while.

Mawson has taken a dislike to Error after he complained bitterly about how little time the men have to themselves.

'We're expected to work seven days a week. When are we supposed to relax?'

'Relax?' Mawson could scarcely believe his ears. Surely no one on the expedition came to Antarctica to relax. Such talk doesn't just disappoint him; it makes him angry.

Error was originally brought on as the medical officer for the second party, but when Mawson consolidated two teams, he had no need for two doctors. Error's only duty now is to replenish their ice supplies for melting over the stove. They use lots of water. It is undeniably the worst job. Heading outside in all weather with an icepick, chipping away while battling the wind until there's a full pail. A prisoner breaking rocks has an easier time.

Mawson thinks back to the letter he received from Error's mother when he offered him a place. *Don't take him, he's my only son*, she had said, claiming his health was poor and he might perish because his system was weak. It was as pleading as the letter Hurley's mother had written. But like Hurley, Error had passed the health check. And with flying colours. If only Mawson had taken more notice of Error's mother than the medical examination, he would have refused to take him. How he regrets that now.

Despite the discussion taking place out of sight in Mawson's office, the terse words exchanged between Error and Mawson were overheard by everyone in the hut. Ninnis, who records everything in his diary, listened intently. His bunk is right outside the curtain that passes for a door. Perhaps that is why he is called upon to do Error's messman duties the following day – a punishment for eavesdropping? He could hardly help it.

A messman's duties are far from glamorous – he's scarcely more than a slave for the day – but at least they can be performed indoors. They all take turns, and every eighteen days it comes around. Up at 7 a.m., the messman dishes up the food, sweeps out the hut, melts down ice, fetches coal, empties the ash bucket and the rubbish, and helps whoever is on the cooking roster. But his job is not done until the evening's washing up is completed. That is a monumental task, with eighteen men in the hut, each with two dirty plates, a bowl and a full set of cutlery. A hundred and eight elements get cleared from the table, rinsed and clunked in soapy boiling water along with all the dirty pots from the kitchen, and the chisel and mallet that the cook uses to break up frozen meat.

Ninnis never thought he would enjoy doing the washing up. But he loves it when it is his turn. This evening the hut is particularly freezing. He relishes an opportunity to sink his hands in the warm dishwater. He cares little that the water is murky after all that has passed through it; the warmth is an utter luxury. How will they wash on their sledging journeys, he wonders. How will they go to the toilet, how will they eat? There are so many questions. He doesn't consider the most basic question of all – how will they survive out there? And yet he should.

CHAPTER EIGHT

'Hey, have you seen what day it is today?' calls Ninnis from his bunk. 'It's the twenty-ninth of February.'

'A leap year!'

'What is *leap year*?' Mertz pipes up in his heavily accented English.

'A year when there is an extra day in February.'

'What is it called in Swiss German?' asks Madigan.

'*Schaltjahr.*'

'*Schaltjahr.*' Madigan tries it out. 'Did you know it is traditionally the only day when young ladies can ask for a man's hand in marriage?'

'Relax, lads,' shouts Hurley. 'We're safe down here!'

'You would be safe anywhere!' Bick shouts back.

There's general laughter and other comments called from different corners, all of them rude and intended to be taken with good humour.

Again they're confined to the hut by poor weather. Mawson has considered it prudent to postpone the sledging excursions, and the men are getting restless, with nothing to do but wait. Blizzards have dumped drifts several feet deep, and the winds have chased even more snow down to the coast from the interior. There is an endless supply. The verandah is full and needs to be cleared, the rocks are covered, the walls of the hut obscured. The dogs need no encouragement to find shelter. They hunker down behind whatever barrier they can find and soon disappear under an insulating blanket of snow. A few times, when venturing out to check on the scientific instruments in the little hut where they take readings in the earth's magnetic field, Madigan has stumbled right over the top of a dog without even realising. A whine, a yelp, a fuzzy snow-covered head emerges.

During a brief lull in the weather, Ninnis put on a show of expertise, harnessing up the dogs Ginger and Shackleton, two of the best pullers, with Bick and Madigan sitting on the sledge. They do a quick tour of their immediate surroundings, where they count 146 seals, who are certainly not bothered by Cape Denison's unpredictable micro-climate. Then the weather closes in, the wind picks up to hurricane strength and everyone beats a hasty retreat to the hut. Playtime is over.

'Back indoors, lads!'

With men like Hurley, there is never a dull moment in the hut. His small darkroom tucked behind the coal range is often the centre of high jinks. While it is technically a creative refuge where he can develop his photos, others often disappear in there when they are planning culinary antics or surprises that they don't want curious eyes to see.

The shows that they sometimes put on for each other in the evening require an off-stage area from which actors can burst for dramatic effect. The biggest laughs are reserved for the men who go into Hurley's dark room as men and emerge as young maidens, transformed by wigs and fancy dress. 'The Society for the Prevention of the Blues' is what they call their regular cabaret.

When the gramophone records get a little stale, the hut band can always be convinced to strike up some tunes. Correll has his piccolo, Ninnis has an accordion and a mouth organ, which he lends to Bick. Hunter has fashioned a kettle drum from a large tin, Laseron has a comb that he plucks and tweaks to great effect. Alfie can always be depended on to provide further percussive pizzazz using a variety of workshop tools and an assortment of lids, while

Hurley is always keen to further bolster the wind instrument section by humming down a length of rubber tube while employing the gramophone horn as a particularly effective means of amplification. It's all great fun. But it is really not why they have come to Antarctica.

Mawson is starting to feel a sickening sense that this could all be a dreadful waste of time and money. The main purpose of the expedition is to map as much of the coastline as possible. He wants passionately to hoist the Australian flag over this whole region. This can only be done by getting out and about. Everything depends on the weather, and the weather in Cape Denison so far is proving an unmitigated disaster. He wants to start on their sledging program, or at least start preparing the men to be active and productive members of the expedition.

Every man has been allocated winter clothing and sledging equipment. Mawson spent the afternoon demonstrating how the clothing can be used to best effect. Layering is critical. Each man wears a wool singlet next to the skin, then woollen long-johns provide a further thermal layer. Next, camel-hair trousers and a thick woollen jumper. Balaclavas and two pairs of socks keep the extremities protected. Mitts made of reindeer or wolfskin are worn over the top of gloves to guard against frostbite in the part of the body most vulnerable to exposure – the fingers.

The Burberry overclothes and hood are made from waxed cotton and provide a vitally important windproof layer that extends over the entire body.

'Important not to overdress. You'll overheat. As soon as you stop, the sweat will freeze against your skin and you'll get chilled. Better to wear something lighter while on the move; you'll stay warm while your body works. You can layer up when you stop.' Mawson looks around the assembled men. There are no wisecracks or laughter as they inspect their kit. It is clear that they respect Mawson's experience. They would be fools not to. After all, this is a man who has trekked over 2000 kilometres in pursuit of the magnetic pole – a point on the planet that doesn't even exist except in magnetic terms.

A few of the men slip their feet into the fur booties called fineskoes to try them on for size. They are immediately taken by how warm they are.

'These will remain supple even in the coldest conditions,' says Mawson. 'Your leather boots won't do here. They will offer no warmth at all. They'll freeze rigid. They might even freeze to your feet.'

It feels good to share his hard-won knowledge with the men. His thoughts hark back to the cheering crowds of well-wishers in Hobart, the flotilla of boats that followed the *Aurora*'s progress down the Derwent estuary. Everybody is expecting great things from the expedition. He hopes they'll have something of value to share with them. A survey of the coastline, invaluable geological knowledge, magnetic work – he might even send some men to chase the magnetic pole.

NIMROD

1909 – THE CREVASSE

It's evening. The sledge has been unpacked, the camp set up. A light breeze is blowing. The Prof is wandering close by, his head raised skyward as he enjoys the soft light on the underside of the clouds. He is murmuring some lines of poetry as though trying to remember the exact wording.

Inside the tent, Mawson has finished arranging things and has now turned his attention to the day's photographic work. Suddenly he hears the Prof call.

'Mawson! Are you busy?'

'Well, yes, Professor, I am at the moment.'

'Mawson, could you spare a few minutes now?' His voice trails off. Mawson can hear the note of hesitation in his voice.

But he calls back, 'I'm just developing some films, give me a minute or two.'

The Prof often has interesting things to say or point out. He is generally a very pleasant travelling companion out in the field, even if he doesn't always pull his weight – when they make or break camp and while sledging. But he is twice Mawson's age – at fifty he is certainly not as robust as Mawson. His voice pipes up again. This time it sounds rather more plaintive.

'Mawson! I'm very sorry! I don't want to trouble you! But I don't think I can hold on any longer, I've fallen into a crevasse.'

Mawson scrambles to his feet and flings open the tent. The Prof whimpers; his hands are barely visible as they slide over the lip. Mawson lunges for him and grabs his sledging harness. Hauling him up and onto the flat, churned-up snow beside him, Mawson feels adrenaline surge and radiate into his fingers and toes. The Prof is panting, either from shock or effort or cold. His legs and torso are caked in snow. Clearly he had plummeted through the surface, having been caught out by a collapsing snow bridge.

'Thank you,' he says, catching his breath. 'I was beginning to wonder . . . if I was done for.'

Mawson lets out a sigh of relief. 'You know, you could have yelled. I'd have come running.'

Polite to the point of ridiculous, the Prof is as eccentric as they come. How soundlessly he was swallowed by the crevasse. It could easily have been Mawson himself. A crevasse doesn't discriminate. Old, young, fit or faint. Anyone can go down. And if it's deep enough, they don't come back out again.

CHAPTER NINE

It is a meagre but symbolic first step. A flagstaff has been erected six kilometres inland. It's the first of many markers they will place in line with the hut, to mark the way back down. When visibility is low, any help in getting home safely will be crucial. What they can see of the interior to the south is a featureless expanse that extends all the way to the blur where the land ends and the sky begins. Since mid-February they have waited for clear weather. They have certainly had clear days. But the wind has a habit of spoiling everything. Now it is the first day of March and winter won't be far off. Will the wind get worse? Is it possible for it to get worse?

Otherwise, they're in good shape. Hut and workshop are finished, covered verandahs are stocked with food, dogs are healthy and happy, scientific equipment has largely been set up and they will start taking readings at the same time every day – temperature and wind speed.

The wind often assails Observation Hill, where Madigan has installed a couple of weather screens – the louvred boxes housing the weather instruments. The only problem for the expedition meteorologist is that these need to be checked several times a day, which can be a mighty struggle if the wind is up. Others are on hand and keen to help, although they are guaranteed to suffer in new and unusual ways for their efforts. Most recently, Ninnis discovered an odd sensation when he tried to blink and found that his eyelids had frozen.

Ninnis writes in his diary that the one consolation to climbing up Observation Hill is the journey down. A white-knuckle ride sliding on one's bottom. There is quite the competition to see who can get down fastest. On numerous occasions Ninnis declares to his diary that he will stop engaging in such antics – his winter trousers are starting to wear thin – but he can't help himself. It's too much fun.

The wireless receiving equipment has been set up in the corner of the workshop. Bouncing Billy, the wireless operator, has everything connected and tested and ready to go. They are all impatient to be able to exchange messages with the outside world. All they need is for the wireless masts to be erected. But the wind still prevents it. If only they could be assured of a few days of settled weather.

There has been a bit of anxiety around the need to move the acetylene generator that they use to light the hut's interior. It has frozen solid outside and has ceased to function. If they're to have light in the hut, they'll need to move the whole assembly inside.

Housing such volatile equipment in their midst is not ideal, but it's the only place warm enough for it to work. They spend a fair bit of time rigging up a platform under the rafters to keep it out of the way. The mechanically minded among them - Bick and Correll - tinker for a long time, and the equipment is eventually restored to good working order. Tubing snakes across the ceiling and down into the various burners that light their workspace. With only small skylights and no windows, the hut is very dark indeed. Being able to see what you are doing is rather important, particularly in the workshop space. Given their greed for light, they are mostly prepared to turn a blind eye to the dangers. Not so John Close.

'Run the generator in our living quarters? What! We could all be blown sky high.'

Close is developing a name for himself as the hut's resident pessimist. His negativity extends to most things: *The ship will never return and we'll be stuck here forever. The wind will blow our hut down and we'll all freeze to death.* They have all tried in vain to reassure him about his dire predictions. Now he is the butt of many good-humoured jokes. Madigan started calling him Terribus for his attitude of doom and gloom, his complaints at mealtimes and his general fussiness. It's not just food. No crockery or cutlery is up to scratch.

'This plate is too small' he cries on sitting down.

'I can't eat this, my knife is blunt!'

His constant grousing about portions means he is the focus of tabletop teasing with the messman either dishing up his breakfast bowl to overflowing or offering barely a scrape of porridge, while he is given an immense wooden stirrer as a spoon. The acetylene generator has him in fits of unease. Already there are plans afoot to prank him.

Eager for the least sign of progress, Mawson announces their first foray. It won't be too ambitious or far-reaching, but some of them will be out for up to a month. Four sledging parties will set out to see what kind of terrain lies beyond their coastal enclave. When every member of the expedition is a complete novice, obtaining any kind of experience before they start their main surveying journeys in spring is vitally important.

Getting onto the plateau will require pathfinding up the glacier ramp. Madigan has already fallen into a crevasse while taking a walk over ground that looked perfectly stable. Luckily he was with others who were able to help when his foot unexpectedly disappeared and he slipped in up to his shoulders, which caught on firmer snow, even as his lower half dangled in empty space below the surface. Skis would have provided some measure of protection. But they need to get better at managing on skis too.

Sledges are loaded with equipment and fitted with masts and crossbeams so that sails can be unfurled. Although they might be fighting the wind when hauling up the steep incline that leads south from Cape Denison onto the plateau. Sledging provisions are divvied up into ration bags that Mawson's fiancée Paquita and her mother have stitched. The men have taken to calling them Paquita Bags. Pemmican will be the mainstay of their sledging diet, dried meat ground up with fat. A dark lump, it does not look appealing in the least, but melted down with snow to form a stew, it will be met with grateful gulps. Or so Mawson assures the others.

The weather is still their only impediment. Will it hinder their progress? And the wind? Will it harass them as viciously inland as it has on the coast? The anemometer measuring wind speed registered 122 kilometres per hour at the end of February. Had the centimetres-thick steel shaft of the anemometer not snapped, they might have seen that the wind was even stronger.

Any job is ten times tougher in its blast. Even walking takes a superhuman effort. Pitched forward on a ridiculous angle, any man caught in the gusts must throw his weight into each stride. Then, just as he heaves himself at it, the gust that was holding him up whips away, leaving him unsupported. He falls straight to the ground. To add insult to injury, the wind blows him along the surface a metre or two, leaving him bruised, dispirited and wondering why the blazes he's here.

Madigan has more or less declared a world record.

'Cape Denison is the windiest place on earth,' he says.

It's hardly a cause for celebration.

CHAPTER TEN

Ninnis and Mertz sit in the completed workshop, a pile of tent poles between them. It's just like peeling potatoes on the ship, except they're out of the weather.

How pleasant to be engaged in an activity indoors and not have to fight the whirling devilry outside, although the noise of it fills the workshop. Somebody said it hit a world record of 128 kilometres an hour yesterday. It has outstripped the Beaufort Scale.

Their task is tedious – filing down the lumpy bits on each bamboo pole. While they were sleeping in the tents during the construction of the hut, holes were worn in the fabric – with all the flapping in the wind against these tiny rough nodes. When they are sledging, they don't want to suddenly find their tent more sieve than shelter.

Mawson is outside checking on progress at the magnetic hut. It's vital that the instrumentation is fitted before the sledging starts. The framing that Webb and others painstakingly erected was blown down in the last blizzard. Not to be outdone, they have put the framing up again, the sides lined with the same waterproofing layer they used on the roof of the hut, but then another layer of hessian bags and finally sheepskins are added. For extra protection, several tonnes of rock are piled around the base. It looks very funny. Not remotely like a scientific laboratory - more like a primitive hovel where a penniless prophet or an ancient beggar might live. The interior of the hut must be kept at a constant temperature if their results are to be accurate.

Throwing himself behind the effort of hefting rocks from the moraine to the hut, Mawson was fascinated to find traces of copper and gold. There are emeralds too; a particularly large one embedded in a rock that Mawson will take back to his office to examine. His brief geological discoveries cause excitement; what other precious metals and gems are lurking in the rock of Cape Denison. Inevitably there's talk of getting rich. Bickerton tells of his brief career as a treasure hunter. Mawson ignores it all. Like world records, glittering treasure has no hold on his scientist mind.

Mertz is also hard at work, but stops when one of the dogs approaches him, shivering and miserable. It's the one called Caruso, named after a famous opera singer. He is in a fearful state, whimpering, his fur matted with blood. Mertz calls Mawson over and together they manhandle the wretched animal back to the hut and into the workshop where Ninnis and Madigan are working on the tent poles. Mawson calls Dr McLean to bring his medical kit.

On closer inspection, Mertz finds a tarred string has been caught around the dog's neck and pulled tight like a noose. It has cut deep into his flesh – Caruso yowls when Mawson pulls the icicle-encrusted fur away from the wound.

'Good grief!'

'It's gone right to the bone in several places,' says the doctor. 'And the flesh has frozen. He'll be in a lot of pain when it unfreezes.'

As if on cue, Caruso starts to let out long sorrowful cries.

'We'd better knock him out with some ether.'

'How on earth did the string get there?' Ninnis asks when the dog has been sedated. His distress is obvious as the doctor trims away the fur so he can stitch up the cut.

The doctor pauses in his surgery to scrutinise the red patches on the dog's skin, all too apparent now his fur has been cut back aggressively, 'Oh dear, looks like he has mange too.'

'Mange? What is mange?' asks Mertz peering over Ninnis's shoulder.

'Parasites under the skin,' says Ninnis. 'Little mites. It's highly contagious.' Seeing Mertz's blank look at the word, he clarifies, 'The mites move from dog to dog. I wonder how many others are affected.'

'Even so cold! How can they survive?'

'The dogs or the mites?' says Ninnis.

'It demonstrates how well insulated they are with their fur,' marvels the doctor. 'The snow gathers on the surface.'

Ninnis has seen otherwise. 'Unless they're exposed to the wind. It drives snow up under their coats and it thaws next to their skin only to freeze again. Quite a few have iced up. When I went to fetch her for sledging, Ginger was stuck to the packing cases she was sheltering behind.'

'We will make him a little bed out of some of the packing straw and we'll need to tie up his legs,' McLean tidies up his medical kit, 'to keep him from ripping out these stitches.'

Hurley is in his darkroom, having shot some wildlife portraits during a fleeting period of calm in the morning. While he was out, the wind had picked up, and Hurley's excitement grew when he saw the effect it had on the waves. Soon the entire scene had disappeared in dense sea spray, which was thrown 30 metres into the air to then freeze immediately into ice chips that were flung onto every surface, including Hurley and his camera. He realised he could not stay out for long on the icy slope that falls away into the sea. Hurley admitted defeat and headed back to the hut, crawling on his hands and knees. The wind can easily lift a man off the ground when he is upright.

Despite the perils of operating his equipment with bare hands and a painfully exposed face, Hurley is determined to capture the wind. Stills convey some of its force when a man is in the frame, but moving images are best for expressing its strength and unrelenting nature. A number of times he has set up a sort of shelter made from blocks of hard-packed snow. Kneeling behind it and out of the wind, he has been able to shoot the men attempting to go about their business in the blizzard. Their bent forms struggle against the invisible force as snow courses past them to head height in horizontal icy ribbons. On such occasions, Hurley's hands must take the punishment. It is impossible to crank the handle of his cinema camera with mitts on. As they start to burn with white pain, he grits his teeth and thinks of his art. Nobody has ever obtained such astounding and original images; nobody would be that mad.

Just before dinner is served, Hurley reappears with a contraption that looks like a telescopic lens on a tripod.

'Transit of Venus tonight, lads,' he says.

'You can't be expecting to go outside to do your astronomy just before dinner,' says Bickerton, who is already setting the table.

Hurley ignores him and sets about adjusting his focus as he peers into the eyepiece. 'Ah yes,' he says. 'Got it.'

Realising it is one of Hurley's jokes, a few curious men line up.

'All right, what are you playing at?'

One by one they take a look and give a chuckle at the enormous image that has been terrifically magnified.

'Hey!' says Webb, suddenly indignant when he looks through and sees that Hurley has trained his telescope on the wall of his bunk. 'That's my girlfriend.'

'Yes, like I said, the Transit of Venus,' says Hurley triumphantly. 'Isn't she a beauty!'

'Clear your things. You've had your fun,' says Bick, knowing full well that for Frank Hurley the fun is never over. And how glad they are for it

It's a Thursday, so that means penguin for dinner. They have it twice a week now. They have seal twice a week too – always on Tuesdays and Fridays. Canned meat on Wednesdays. On Sundays they are treated to mutton. They have been busy stocking their larder with local wildlife. The mutton and penguin is stored in an improvised meat locker under the hut foundations. It's one of the few places safe from the hungry dogs.

After dinner is finished, Mawson stands to address the men on sledging outfits. Their first sledging foray has been considerably delayed. By now at least they all understand from bitter experience the necessity of wearing appropriate clothing.

'Our plans may have been thwarted by weather but I still hope we'll get to do some sledging, head inland in the next week.'

'The condition of the dogs . . .' someone starts to say.

'No, we're not using the dogs,' Mawson says. 'We can't afford to lose any. This is really just a practice. We'll save them until the real work happens in the spring.'

Chatter starts up. Mawson holds up his hand for quiet. 'I wanted to say something.'

The men around the table fall silent.

'I wanted to say that I know the hut has been fearfully cold.'

There's a murmur of agreement. They manage to keep the interior temperature above freezing now, but it has never exceeded 7 degrees.

Someone calls out, 'Especially when idiots fall asleep and fail to stoke the fire with blubber . . .'

'Hey!' Mawson holds up his hands for quiet again.

The guilty nightwatchman folds his arms across his chest defensively and scowls.

'All I want to say is, don't be tempted to use the special clothing you've been issued. Any old clothes will do about the hut. You don't need to wear the special clothing – that is for our sledging journeys. It's never as warm once it's been washed. Save it. We have to start our sledging in new clothing. That's the real work of the expedition.'

'I thought the real work was now getting rich,' says Hurley. 'Shouldn't we all be mining for precious gems?'

'Sorry, Hurley, if you are referring to this morning's geological discoveries, I can lay your mind at rest. The large emerald turns out to be something called green apatite.'

'Green apatite!' he scoffs. 'Got no appetite for that.'

CHAPTER ELEVEN

A phenomenon they are all fascinated by is the St Elmo's fire that flares up from time to time, its unearthly purple glow tracing the outline of certain objects outside. The nightwatchman on his rounds often sees it play about the wires of the chimney stays on the roof, or illuminating the scientific instruments with little balls of what looks like blue fire. Another favoured spot is the weather vane on top of the weather screen where the meteorological instruments are housed. One calm evening while out for a walk, Madigan discovers his ice axe is a marvellous conduit for St Elmo too. His excited shouting starts a stampede of other enthusiasts who rush to the same spot and chop at the air, hoping to possess a bit of electrostatic excitement themselves for a time.

Another visual spectacle is the nightly Aurora Australis, or Southern Lights, which can be seen as a golden rippling curtain or flowing tendrils of green silk whenever the sky is clear of cloud. Now that evenings are growing darker, the aurora emits enough eerie light to make the waves in the bay glisten. Noting down the frequency and intensity of the auroral displays is part of the meteorological observations that the nightwatchman must make at regular intervals during his lonely vigil. Even on overcast nights, even during a blizzard, whoever is taking the night watch is expected to venture outside and accurately record conditions. The aurora is now known as 'Watchman's Terror', because it's a devilishly unpleasant job that requires kitting up in multiple layers and braving the hurricane that always seems to be raging outside – all in the interests of science.

With their collection of instruments, Mawson and his men are gradually piecing together a scientific understanding of their environment. There is the magnetograph to measure variations in the earth's geomagnetic field; the barograph to keep track of air pressure; the tide meter to monitor sea levels; the snow-drift gauge improvised by Alfie and Madigan from a stove pipe; the thermograph to record variations in temperature; the nephoscope for measuring the altitude, direction, and velocity of clouds;

and the electroscope for measuring an electrostatic charge in atmosphere.

Then there is the poor overworked anemometer to gauge windspeeds; 'Annie' is the name they have adopted for it. Annie can certainly offer proof of how inhospitable Cape Denison has turned out to be, even if she cannot offer answers to such questions as: Where is it all coming from? Where is it rushing to? Why does it want to wipe them and their hut off the face of the earth? Annie has no idea. There's also the puff anemometer to measure the short, sudden wind gusts – this is called the Puff-Annie.

Not surprisingly, the anemometer up on Observatory Hill is getting a hammering and requires frequent repairs. Winds that frequently exceed hurricane strength, dense drift, blizzards – it is barely mid-March and they wonder: If this is autumn weather, what will winter conditions be like? They can but wait with bated breath and dripping noses.

Madigan often waits for ten o'clock to take a wind reading. Bending his head low to shelter his ice-encrusted face from the wind, and breathing heavily through his nose to circulate warm air around his face and prevent his eyelashes freezing, he feels a drip of thawed snow run off the tip of his nose. After a moment or two a stalactite forms there. Further drips gather and some drop to his frozen sleeve, where they immediately turn to ice. A stalagmite forms. It's a pleasing little tableau, but he has no desire to watch the phenomenon until the mite and tite make icy contact. Staying stock still, he might freeze himself before then.

One Sunday morning, the very last day in March when the wind is mild, Madigan and Alfie head up to Observation Hill to change the records in the meteorological screen. Bick is also following, although some distance behind. By the time Madigan and Alfie reach the base of the hill, a stiff breeze is building and tossing drift several feet in the air. Before long, visibility is limited to barely an outstretched arm. They call to Bick, but there is no point, as the clamour of the wind whisks away their words. Coming like a flash flood at knee height, a sudden gust tackles Madigan and tears the tin containing the records from his hands. He gropes after it and Alfie watches, bemused, as Madigan's figure blurs and disappears amid the spinning white drift.

When Madigan returns to the spot empty-handed, Alfie is no longer there. Obviously he has continued up the slope with Bick. But on reaching the Stevenson screen, nobody is to be seen. Preferring not to be out in such conditions, both Alfie and Bick must have turned back to the hut. How wise, thinks Madigan, hurriedly retrieving the records and making a break for home in the increasingly heavy drift.

On regaining the hut, Madigan is shocked to see only Bick is back.

'What have you done with Alfie?' Madigan asks.

'He was with you. I turned back.'

'Oh, Bick.' Madigan stops and grows fretful. 'He wasn't.' He decides against removing his outdoor clothing. Instead he calls through the door into the main hut.

Mawson, Hurley, Ninnis, Bage, Webb and Dr McLean kit up and join Madigan and Bick. It's the last thing they feel like doing, but heading out into the clamour is entirely necessary. With the hut so close to the coast, Alfie could easily find himself on one of the steep icy slopes that fall rather precipitously into the sea. With such wind a constant, the sea never gets a chance to freeze over. To fall or slide into the water without companions to come to the rescue, a man would face a certain and rapid death.

Meanwhile, Alfie is not lost. He is sheltering behind a rock and loosening his trousers, which have filled with snow after he repeatedly lost his balance. He had made it to the Stevenson screen on the hill, but when the others didn't show up, he went down again. Somehow he had lost his way and ended up in a locale on the flat that he did not recognise. So he climbed back up the hill and tried again, coming down what he hoped was the regular route, although nothing could be seen of it. His field of view is obscured by drift. He might as well be blind. And the howl of the wind is so loud, he might as well be deaf and dumb, for there is no hope of hearing or being heard. He is more lost than ever. But not yet afraid.

'Alfie!'

'Hodgie!'

'Alf!'

But Alfred Hodgeman remains ignorant of their calls. The mitt he removed to better adjust his trousers has blown away. He curses his stupidity. His gloved hand feels the immediate pinch of cold, his fingertips already sparking with frost nip. He gets up, holds his head down and tries to think what direction the wind was blowing from when they left the hut. No idea. Just as well, it could easily have changed its mind and swung around to attack Cape Denison from another angle.

Alfie closes his eyes and imagines Cape Denison on a good day. The hill, the hut, the little lake, the ridges that rise on either side, the sea, the slopes down to the water's edge, the Boat Harbour, the glacier leading up to the plateau, the moraine where they gathered rocks to place around the magnetic hut. He must steer clear of the slopes.

Making headway against the wind is exhausting. He crawls for a time. Then slides on his bottom, shifting forward in increments. His beard has iced up, and lucky too. It acts as a sort of shield against the windblown snow that lashes his Burberrys with a hissing sound that becomes quite hypnotic when he's making such slow, deliberate progress along the ground. Is it several hours or is it ten minutes that he has been sliding across this ice? Time has ceased to have meaning. A dark shape floats into view, then another – a cluster of rocks like islands on a vast swirling sea. How happy it makes Alfie to rest his eyes on something solid and real. Then there is a road of rock. It is the moraine! He must be close. He gets to his feet, braces against a gust, then walks stiffly but resolutely towards where he hopes to find safety. There – it emerges like a watermark – the hut. He is stiff with cold and suddenly flooded with gratitude. Exhausted, he falls through the door. He's been lost for three hours. How easily he could have been lost for ever.

NIMROD

1909 – THE MAGNETIC SOUTH POLE

It is the first week of October. Shackleton has put the Prof in charge of the sledging party consisting of Mawson and Dr Mackay. They are to locate the Magnetic South Pole, or at least attempt to. Crossing the sea ice of McMurdo, then initially skirting the coast, their route leads many hundreds of kilometres inland. Through uncharted terrain they will search for this moving target, shifting with the earth's magnetic field. Before 1902 it was going east; now it seems to be going north-west.

'Tally-ho!' says the Prof, undaunted. 'We will give up all for the magnetic pole.'

A tonne of equipment and food spread over two sledges

is a horrible load to haul. Progress is slow; relaying triples their distance. Drag one forward, double back, collect the second. It takes them a month to travel a distance they might walk in three or four days. Surely better to focus on geology?

'The Prof is certainly a fine example of a man for his age,' Mawson confides to his diary. 'but he is a great drag on our progress.'

Both Mackay and Mawson wonder if the Prof is pulling as they are. It appears that he is simply leaning his weight into the harness. The way he tucks his thumbs into his braces would suggest that he is enjoying a stroll in the countryside. Yet all around them are perils – jagged pinnacles and gaping chasms, flaws in the ice and treacherous gashes hidden under deceptive layers of snow. Dumping one sledge and provisions at Drygalski Glacier, they leave letters of farewell in a biscuit tin. Just in case. Proceeding inland, they can but pray for a favourable outcome.

'I'm going!' Mawson yells as the sledge careens after him down a crevasse, pulling both Mackay and the Prof off their feet. Mawson swings on the end of his rope, which cuts into the surface like a knife. By some miracle the sledge wedges crossways; a dramatic halt, a lucky escape.

Their route is altered. A blizzard blows in. Still they persist. It can't be far. Across glaciers, hauling, pushing. By New Year's Day they are exhausted, hungrier than ever, constipated, by turns snow blind, fearful that their tent is too fragile to last. Frostbite has them all in its nasty embrace: Mawson's cheek, the Prof's nose and

Mackay's hands are affected. Mackay's spirits could not be lower. The Prof is morose. By Mawson's calculations, they are still far from their goal.

They lurch on for one more day. Hoist a flag and declare a victory of sorts. They can't be entirely sure that they're in the right spot. But it's close enough. The Prof takes possession of the region on behalf of the British Empire. A photo. Three cheers for the king. Then they return to the coast, following their outbound tracks across the monochrome landscape. Yet the Prof is not well. The strain has been too great. His boots have frozen to his feet. He stumbles and mumbles. Dr Mackay believes he is both physically and mentally unfit to be their leader.

CHAPTER TWELVE

Hurley is the day's cook. He brings the mutton to the table singing a ditty he and Hunter have composed about Alfie getting lost in the blizzard. It rouses everyone to song and general merrymaking. Usually Sunday is the day they sing hymns, but Alfie's disappearance meant that they did not hold their usual church service that morning.

After the meal is cleared away and as the men begin to retire to their evening pursuits, Mawson clears his throat and calls for their attention.

'Nobody needs to be told how lucky we all were today. And not just you, Alfie. Every man that went out into the storm to look for you was taking his own life in his hands. We had Hurley roped up, dangling over the sea doing sweeps of the coastline, we had Madigan and Bick out, Bage and Webb, McLean, and Ninnis, whose face was blasted . . .' he pauses. 'I could have lost any number of you. I could have lost our doctor. As it happens we didn't, but a number of you are suffering the effects of frostbite. And frankly, I'm furious. We can't have men charging around out there in poor visibility. Madigan has already had a fall down a crevasse. We just can't take unnecessary risks.'

'Hard to see whose fault this was, though,' says Madigan. 'Alfie shouldn't take the blame.'

Mawson knows it; they all do: they lack experience. These young men simply can't read the conditions. And they have yet to develop a healthy respect for the Antarctic environment. It's all been a bit of a lark up until now.

'Perhaps it would have been best to rope up,' says Bick.

'Assess the risks,' says Mawson dryly. 'Think things through. Take it gently. We can't have you adjusting trousers in a storm or chasing after mitts or gloves or whatever out on the plateau or on the sea ice.'

'It was the tin for the weather records that blew out of my hand,' says Madigan. 'I had to run after it.' But his redundant comment does not get a response.

'I think he was talking about me, Madi,' murmurs Alfie, cradling his frostbitten hand.

Mawson continues in his sternest voice. 'I need you all to take responsibility for your safety. And it starts with your kit. Check that it isn't loose. Modify it to fit your body. There's plenty of candlewick – use it to make a harness so you can secure your mitts. Those who must venture outside on nightwatch, please remember to sign in and out in the logbook, so if you haven't returned, we can see how long you have been gone.'

Lying in his bunk in Hyde Park Corner, the quaintly named corner of the hut he shares with Madigan, Bick and Ninnis, Mertz is cross. He would have loved to be part of the search and rescue. 'You must have woken me from sleep,' he says plaintively.

'You *should* have woken me up,' corrects Madigan. He's a bit moody after the stern words from Mawson. Madigan is not entirely fond of him or his serious demeanour, which he sees more as arrogance or priggishness.

Mertz continues, 'Such excitement outside! And I was sleeping. The whole time.'

'You don't wake the nightwatchman after he's finished his watch,' says Bickerton. 'You let the poor blighter sleep!' he gives Mertz's bunk a playful kick.

Madigan does the same to Ninnis. Except a little harder than Bick. Ninnis's planks shoot up and send the young man and his thin mattress into the air. He protests but does not retaliate. Normally such an action would result in an immediate escalation or even an all-out war.

But Ninnis is quieter than usual. Sunday is the day when he is most likely to feel homesickness. Not in a melancholy sense, exactly. But he is looking forward to returning to the familiar. Generally it's the hymns that set off such reflections. Madigan is heartened that their longitude does not differ greatly – in fact only 29 seconds – from Adelaide, and he can at any moment imagine what his loved ones are up to.

But Ninnis's world is upside down. He misses his dad. He misses England, the traffic and bustle of London, restaurants, theatres, reading newspapers; but also the country lanes, fields, villages. He can imagine the trees would now be in bud, the primroses should be coming out. The music brings it all back – how cut off they are from the world and their loved ones, who will no doubt be picturing scenes of carefree autumn sledging, not hurricane winds and a hut that is slowly but surely disappearing under a colossal snow drift.

The day's events have also rekindled Ninnis's fears. Amid all the excitement of arrival and discovery and setting up winter quarters, he has pushed a gnawing sense of trepidation from his mind. And now it has returned. He flicks back in his diary to an entry he made almost exactly two years ago, when he was first selected for the expedition.

He could have written it yesterday. The memories come flooding back.

How honoured he had been to be assured of a place, one of the few Englishmen chosen to take part on an Australian expedition. And after two failed attempts to depart with British expeditions led by Shackleton and Scott, he savours the success with the AAE as a particularly sweet triumph.

To share the proud news of his appointment with his dear father had been one of the most joyous occasions of his life. His father, who had not only been a polar explorer himself but who had also acted in the service of the Australian government in surveying in the outback. It seemed a pleasing symmetry that he was keen to point out to Mawson. Perhaps if other, more qualified Englishmen complained of not getting picked, Mawson could refer to the Ninnis family connection with Australian exploration.

Suddenly he had been swept up in all the planning, the negotiations around the purchase of the ship, conversations with Captain Davis, meetings at the expedition offices, lunches and events and talks at the Royal Geographical Society. The intensive field surveying course that Mawson had arranged for him had kept him busy. How seriously he had taken his lessons. He'd even headed out in an English snowstorm to practise with the theodolite, experiencing first-hand the agony of manipulating the screws with bare fingers. That had brought it into stark relief. This was not going to be easy. In fact, it could be fatal. He reads:

Supposing when hundreds of miles from the base or depot I met with a serious accident, say a broken leg . . . What would my companions do? What if they needed to drag me on a sledge? Of course it would slow their pace to such a degree that their food would run out. Being polar men they would surely not leave me.

Should one allow companions to sacrifice their own lives in such a way? Better to do the right thing and end it all.

Suicide under such circumstances is not only excusable but it is the only right course to pursue, he had written.

Reading on, he foresaw other perils of participating in the expedition.

I might go over a crevasse and get badly smashed up, absolutely out of reach of help.

He had resolved then to take a supply of cyanide tablets and keep them on his person at all times. But he had not done it. Such accidents were unimaginable.

CHAPTER THIRTEEN

With snow piled up all around it, the hut is largely protected from the wind. However, it is considered prudent to erect a barricade of boulders and heavy packing cases to the windward side, where the hut bears the brunt of the southerlies. Once this is up, it acts like a bulwark, with the snow collecting behind it. A second line of defence. The roof of the hut has also been reinforced with struts. They're snug but they can't be too smug. If the wind manages to lift a panel or board, the whole roof could tear off in one go.

They have developed a subterranean labyrinth. Tunnels now connect their hut to the outside world, but also serve other purposes too – a nursery for puppies, a slop station for waste liquids from the kitchen, storage for wood. They have unlimited construction materials – snow is not in short supply.

It might feel like hard work if they weren't having so much fun with their picks and shovels. It feels like play, but there is serious intent. The tunnels will enable them to get to their stores without braving the wild weather, without getting fully kitted up, without getting ice-blasted in the face.

Madigan and Bick start a tunnel at one end and by skilful calculations meet Hurley and Dr McLean coming in from the other side.

It's not surprising that this was originally Murphy's idea. Poor Murphy had had quite the task; for weeks he had been digging out the stores and the covered verandah every day. Every day it would fill again with windblown snow so that the only way in or out was to slide in on one's belly, a bit like a seal. It was an endless task, a thankless task, a hopeless task. Murphy versus the Antarctic winter. But he has seized on a better way.

He has also seized on a better way of retrieving the meat from the stash under the hut. It was considered much safer to put it there to keep it well away from the dogs. But with all the snow that has fallen, it's become a very tight space, too tight for a man.

Murphy's latest play is to send a dog under there, and grab it as it tries to escape with a leg of mutton or a frozen penguin. But everything has frozen solid, and a dog cannot pry anything from the frozen ground. The seals and penguins that used to inhabit these shores in their hundreds have left. What reserves they have must be carefully used. Every last bit of blubber and meat – it's a precious commodity now – for dogs and men, for food and fuel. When will their meat supply return? Nobody has any idea, but it's not difficult to understand why they left, with such foul conditions prevailing. Given the inhospitable weather that prevails even in summer, one wonders how creatures would consider making Cape Denison their home at any time.

Any explorations must be made in daylight and with good visibility. The hut can be missed entirely. Mawson himself got lost coming back from the magnetic hut, but eventually turned up alarmed yet unharmed. Recently, while taking his turn as nightwatchman, Correll went outside to fetch coal. He was spun around by the wind, felled and tossed along the ground such a distance that when he righted himself he had no idea where he was. Darkness and thick drift made getting his bearings impossible, guessing his way back to the hut a dangerous enterprise. Remembering Alfie's ordeal, he hunkered down and waited for the dawn. Luckily, on waking Hurley checked the logbook and saw he had not returned. Twenty metres from the hut is not far, yet he might as well have been twenty kilometres from safety, had Hurley not found him. There he was, curled in a ball behind rocks like one of the dogs, but unlike them, he was unprotected by fur.

Frozen rigid, his face obscured behind a mask of snow, compressed into the hood of his windproof jacket, Correll was lucky. Had it happened at the beginning of his watch, while the others slept, he'd have surely died. The temperature is a brisk minus 40.

It spurs another of Mawson's stern warnings. 'Be prepared for every contingency in Antarctica. If the worst does not come, so much the better; if it does, then you are prepared.'

How useful it would be to demonstrate this by getting the men out in their environs during a sledging journey. But the weather will not allow it. Any sledging work keeps getting postponed. When will there ever be a break in the weather? The meteorological records make depressing reading.

Monday – 115 kilometre winds
Tuesday – 96 kilometre winds
Wednesday – 104 kilometre winds
Thursday – 128 kilometre winds
Friday – 160 kilometre winds
Saturday – 156 kilometre winds

Sunday – Annie is broken again. She has lost three vanes, or rather the wind has spun her so violently that they have flown off. And the spindle has snapped. Such savage treatment is becoming standard. Madigan has already cleaned the assembly, soldered the hinges back on. Correll has been hard at work mending her – adding a more heavy-duty steel spindle. How long she'll last is anyone's guess. Mr Fix is rather busy and thoroughly sick of fiddling with the anemometer.

'She's only designed for 30 kilometre winds,' says Madigan as he and Bick heft Annie along the tunnel and out into the fury of yet another evening squall.

It seems cruel to be delivering her again into the gale. Yet this is her job. Seeing the state of the weather, they decide it is too late to venture all the way up the hill. Just as they are about to set Annie down on the stacked coal where it can stay until the morning, a sudden gust throws both men to the ground and lifts Annie – all of her 45 kilograms – up into the air, before dumping her on her head. Broken again.

When they reappear with the mangled anemometer, the cups from three of four vanes gone, Correll cradles his head and moans, 'When will it end?'

NIMROD

1909 – THE PROF

Mawson must assume command. He is physically and mentally strong despite the trials, the hardship. Yet he does not want to. To take over from the Prof who has broken down in body, if not in mind. In the notebook, the Prof delegates his authority, asks Mawson to sign his name.

Four months they've been on the move. Hauling. Freezing. Bickering at intervals. They have run out of time and they're almost out of food. There's no chance that they will get back to their winter base at Cape Royds. Not before the Nimrod returns to pick up Shackleton and the other members of the expedition. At worst they'll need to overwinter on the coast, survive on seals and penguins. What have they got to show for the experience? Mawson knows a better use of their herculean efforts would have been to make a survey of the coastline, not chase an invisible moving target. They should have chosen science over meaningless world firsts.

'Shackleton promised to send Nimrod to look for us on the first of February,' says the Prof with authority.

Mawson and Mackay look at each other; neither are sure they should believe a word he says. And if weather or ice conditions should prevent Nimrod from getting close enough to spy them? Half rations to double the distance they can cover. On the coastline they can replenish supplies from the depot on Drygalski Glacier. Mackay says they're too far north-east, the Prof suggests too far to the south-west.

'No,' says Mawson. 'We're on track.' His navigation skills are unrivalled. He's right.

They soon glimpse the depot. Yet a deep ravine stands between them and their salvation. It completely bars the way. They are so close. The impassable icy gash mocks them. More deviating, false starts, detours and dispiriting falls.

On 4 February Captain Davis makes a second and final sweep of the coastline. The green tent on the snowy outcrop is a relief to see. He sends a signal that he has seen them, then despatches the whaleboat ashore. When they board the ship, the men are nut brown, their frames wizened and worn out, their hands look like talons. But they are alive. They have completed the longest unsupported manhauling journey ever undertaken in Antarctica.

Never again, says Mawson.

This time he's wrong.

CHAPTER FOURTEEN

After four months they are coming to understand that the wind, which they hoped was a temporary or at least a seasonal inconvenience, is in fact a permanent fixture of their environment. They must simply accept it and, as far as possible, work around it with their daily routine now well established.

Correll has been on night watch. He wakes the cook at 7 a.m.. Then at 7.30, as is the habit they've adopted, he places the needle on the gramophone record and music floats out from the speaker, which has been stuffed with a towel so as to not wake the other sleepers too violently. The towel will be wrenched out and the music will rouse any stragglers as the breakfast hour approaches. Any man who has thus far slept through will be given a rather rude awakening just before 8 a.m. with the frenzied jangling of a spoon in an enamel mug and chided with a spirited, 'Rise and shine!' in his ear.

'It's a glorious day,' Correll says to the men ambling towards the table, which he has got the first ones to lay with plates and cutlery. 'Minus 17 degrees, gale force winds, and sheeting snow whipped to the height of a man's thighs.'

In most parts of the world and in most people's estimation, it is a horrid day. But at Cape Denison this is the sort of day the men dream of, a day when outside work will be attempted, although any task that is started may not be completed.

Meanwhile, as soon as breakfast is over, Correll knows he can spend the rest of the day in bed. During his night watch, he took a bath (a pleasure enjoyed once every eighteen days), washed his clothes in the bath water, hung them over the stove, took observations, made bread for the morning, and played with the new puppy, which has been christened - rather appropriately - Blizzard.

First he'll need to rip his blankets away from the wall of the hut, where they regularly get stuck to the ice that forms on the walls. Ninnis has had the opposite problem on the top bunk. Water drips incessantly onto his bedclothes from the rafters over his head. The hut is not kept too warm, but it's still warmer in the interior than it is outside, and as much as they have tried to plug cracks and gaps in the walls and ceiling, water finds a way through. It's dripping on Mawson too.

Wet clothing is also a problem. It doesn't pay to be lazy, to slip from one's outdoor clothing and leave it where it falls, thinking of attending to it later. Any snow left on Burberrys will eventually melt if left in the hut. They have discovered that going out in clothing that is not completely dry is a disaster, as it freezes rigid like a suit of armour. Snow cakes like mad on anything woollen. The hut can look like a jungle of tangled damp layers hanging from the rafters and bunks.

Correll's fingers are incredibly sore and he cannot write as much as he would like in his diary. Screwing the wind vanes on Annie resulted in fearful frostbite, even though he was wearing instrument gloves. His thumb and index finger are particularly bad, and the flesh feels as hard as wood. There's little he can do but hope they'll improve. Doing any fiddly work outside without mitts is a deeply unpleasant experience. Even so, when first leaving the hut, hands freeze in the mitts too. Only when they become encrusted in ice do they provide any decent protection against the wind.

For the first time in months the weather is considered calm enough to erect the wireless masts. It's not going to be an easy job, but if they're lucky and the wind conditions allow, they'll complete the task quickly. The days are shorter now that it gets dark at four o'clock. At just over 30 metres tall, the masts will need to be properly rigged, anchored to the ground with wire stays, and set in place with solid foundations. Somebody will need to be hoisted to the top to do the final adjustments. As this is Hannam's project, he would be the natural choice, but he is too fat to be hoisted skyward. Far safer for him to be seated at his signalling equipment in the workshop than airborne.

Mawson especially wants to start sending and receiving messages – the first communications between Antarctica and Australia. Mawson wants to establish contact with the men on Macquarie Island and they all want to get messages through to their loved ones. Ninnis wants badly to know who made it to the South Pole first – was it Captain Scott or Roald Amundsen.

Madigan and Stillwell had spent the previous day drilling and blasting holes in the frozen gravel with limited success. The dynamite explodes and sends a thin spray of ice and dirt into the air but is scarcely effective when it comes to breaking up the bedrock.

Desperate as they are to establish a link with the outside world after no news for five months, the men all want to lend a hand to the effort on Saturday morning. Webb and Bick, Madigan, Ninnis, Mertz, Laseron, Close, Hunter and even lazy Whetter have turned up. Once positioned on the ground, near the hole that has been blasted into the rock, the first of two lower masts is tilted with the help of some sheer legs acting like a crane. The men haul on the rope, which raises the sheer legs, which lifts the mast into place. Perfectly vertical, it slots into position, whereupon it is fastened with wire guy lines to the surrounding rock.

Buoyed by their success, the men quickly set to, positioning the second mast. Rocks are piled around the base of each mast to ensure they are as secure as possible. If one thing is certain, it's that the winds will return – these must hold up. Bick is hoisted on a bosun's chair all the way up and fixes an aerial to the very top of one mast.

'Watch out!' Bick calls as his hammer falls from his hand and plummets to the ground, narrowly missing two of the eager faces peering up at him.

'Hey! You trying to kill us?'

'Sorry! My hands are bloody frozen,' he calls down.

A cheer rings out over the icy twilight. The moon has risen. Dogs bark. The breeze is picking up again. Time to go inside for the supper that Hodgeman and Murphy have prepared.

Talk over their meal focuses on the imminent communications. Madigan bet Hannam a block of chocolate that they wouldn't be able to erect two masts in one day. He would be sad to hand over his entire chocolate ration if it wasn't such good news.

Hannam refuses. 'It's okay, keep it. We all won today.'

'I made the bet. I will keep my word.' Madigan hands over the precious bounty.

Hannam looks unsure. Then he says, 'I suppose I could share it with everyone that helped make today such a success.'

The others need no persuading and rip into the block of chocolate as they gather around Hannam and his wireless station in the workshop. Hannam starts to tap away in dots and dashes of morse code.

'What are you tapping?'

'Having . . . a . . . hell . . . of . . . a . . . time . . . waiting . . . for . . . calm . . . weather . . . to . . . put . . . up . . . more . . . masts.'

There's not a great deal to see but they are all completely silent as they imagine what messages they would send themselves if given half a chance. After a while, a low murmur passes around the assembled group. Whatever excitement they expected has not eventuated. They realise that whatever is happening can only be heard through Hannam's headset. The men watch his face for clues, hoping for a reply from the relay station at Macquarie Island.

'Hear anything?' asks Mawson hopefully.

Hannam looks up at him with an expression that clearly says *no.*

'It's like watching paint dry,' says Hurley.

One by one the men wander off to pursue other, more satisfying pastimes than watching Bouncing Billy sitting with a look of intense concentration at his station. Late into the evening, he will be listening for dots and dashes – any response. But there is nothing but silence in the ether. It feels as though they are quite alone. The wind howls across the roof. They might as well be the only people in the world.

CHAPTER FIFTEEN

A figure walks into the hut carrying a headless penguin. He is plastered in snow. The wind follows him inside, like an eager dog. For a brief moment it races about excitedly, nipping at his heels and tearing at the wolfskin mitts he has hung by the door. He props the penguin against the wall and stamps his feet, although his appearance does not change greatly. He stands for a few moments, whacking his sides and giving his front a few hard knocks with his fists. He knows from experience that eventually the frozen fabric of his clothing will thaw and regain enough suppleness that he may remove it. But first he must attend to his hood.

A groan escapes from where his face would be, were it visible. But it has disappeared under a casing of wind-hardened snow. His gloved hands tentatively start to crumple away the white crust attaching it to his head. His fingers clear a hole and suddenly he has eyes, frost-rimmed with white spikes that blink twice. Once in relief, then again in frustration at the effort of undressing.

He tugs at the hood and it pulls from his shoulders with a crackle. A head appears, or rather a frost-covered balaclava does. Next he carefully removes his outer windproof layer, still stiff with cold. Now he can finally move from the covered porch to the workshop where it's warmer. Next he'll get busy on the thick ice that has formed around his nose and his mouth. Care is required. Attack the ice too vigorously or remove the balaclava too quickly and he'll rip his own frozen beard out at the roots.

Strange to think it is his own exhalations that hold his head in such a tight icy embrace, the moisture from his breath that has solidified in the air outside, which is well below zero. Once inside the hut again, his own breath will do the opposite – melt this frozen mask. That done, he'll be able to smile and laugh and eat his dinner, even though his lips are cracked, even though his face is numb. Even though he'll have to go outside again to check the instruments – three times a day he does it, even in the foulest conditions. As the expedition meteorologist, Madigan knows that a weather man's work is never done; just like the weather is never done in this land of heavy snows, driving blizzards, ferocious winds and biting cold.

He calls out a greeting as he passes through the workshop, which separates the entry from the main living space of the hut, and where the smells of cooking gather in pockets of warm air. Madigan attracts a few disinterested looks as he slips through the inner door with the frozen penguin tucked under one arm. The snowy comings and goings about their hut no longer offer much excitement. Everyone gets iced up. And they've all seen a penguin before.

Some cheeky individual has hung his socks above the stove and they're dripping into the supper pot. Tonight's cook is Whetter. He grabs the socks and flings them onto the floor before realising that they are his. He raises an eyebrow at the penguin that has landed triumphantly on the cook's table.

'That took a lot of effort,' Madigan says.

'What happened? Where's its head? And where are its feet?'

'Snapped off. I tried to pull it from the penguin stack. Had to use a crowbar.'

Frozen solid. There'll be no filleting it with a knife. Instead, Error takes to the penguin with a chisel and a mallet.

In the background a gramophone record plays a tune that they all know by heart. Some men are lying in open bunks, others slouch at the table over a card game. Madigan plonks himself down beside them while Error complains loudly about something in the oven.

Madigan removes his gloves and wiggles feeling back into his frost-nipped fingers. His hands press to his thawing face. The blood is returning to the surface and his skin burns. He starts tending to his encrusted beard, picking out knobs of ice the size of acorns above his neck cloth. It's a slow business. From time to time he presses his palms to the hard lumps where his cheeks should be and groans. If he had a mirror, he'd see the flesh has turned white. But he's going by feel alone, and his face is a patchwork of frozen chunks.

'Hey, Madi, stop. That's your own frozen cheek you're trying to pull off your face,' says Hurley, looking up from his cards.

Over in the kitchen, the cook's complaining gets fiercer. There's laughter too from others.

'What on earth is *that,* Error?'

'Bread, of course!' says Error, offended.

'It's not. Bread rises. That looks like solid rock.'

'Hey, Boss,' somebody shouts, 'you're a geologist. Give us an opinion.'

Douglas Mawson comes to cast an eye over Error's bread. 'Are you sure you added baking soda? It hasn't risen at all.' He raps on it with his knuckles.

'I'm a doctor, not a baker!' Error protests as others

gather around the cook's table. He shoos them away in annoyance and returns to his pots and pans.

When it's time to eat, the men find that Error's supper is no better than his bread, which he has thrown out the trap door into the snow. He hopes the blizzard will quickly cover it or the wind will blow it away.

'Whetter, this seal liver has the consistency of shoe leather.'

'Blame the seal.'

'I blame the cook.'

'Something is burning . . .' says Mawson. A moment later there's an explosion from the oven.

Whetter leaps from the table. Grabs a cloth. Wrenches open the door to the coal range. Steam billows out and there's a loud fizzing sound then a dull pop as a gooey mess spews from the seam of the tin like green lava.

'Ugh, what the . . .' The label on the overflowing and much-expanded tin has blackened in the fire, but by the oozing contents, it can be guessed that it is a tin of peas.

He is not the only man who has forgotten about a tin defrosting in the oven. They've all done it. Everything arrives from the food store in the verandah solid with ice. He's also not the only terrible cook. There have been others. While they all take turns at the stove, some have proven so unsuited to the task that they have been elected to the 'Crook Cooks Association'.

After the meal, there's unanimous agreement that Error should be appointed President of the Association. They offer him a medal, fashioned from one of the tough little scones he made for lunch. Error gives a small bow. If only it meant he didn't need to cook again he would be truly happy. He freely admits the fried penguin was underdone and oversalted, but he's at a loss to understand how canned food can taste so wrong when all he needs to do is thaw and heat it up.

Overhead the wind growls. The blizzard has raged for four days straight. There's sure to be another one looming. The polar winter advances on their base like a dark tide. There are days when the sun bathes their surroundings in soft light and overhead colours range from lilac to amber. However, even on days of settled skies the wind is the master of all things. It frequently reaches hurricane strength. Ninnis has suggested that all the fiercest weather left over from the creation of Earth has concentrated in Cape Denison. Others say that they were lured to this evil place under false pretences. Nobody can believe that they experienced the week of calm on their arrival. Everyone agrees that the words 'it cannot get any worse' should be avoided at all cost. The rule of this place seems to be that as soon as such words are uttered, a worsening indeed occurs.

PART II

CHAPTER SIXTEEN

Hurley has strayed far from home and he is ecstatic. To be on the loose outdoors and not be blown to bits is a rare treat. His photographic activities have been hampered by the ferocious wind, not to mention the long dark winter of confinement in the hut.

Virtually all the men have been tempted outside today by the promise of another dead calm day, the second one in a row, but the photographer has struck out on his own, eager to make progress while the weather holds. The day is light but the sun has not yet made its appearance. The temperatures are still frightfully low. Not pleasant for men, but Hurley is more interested in the effect minus 30 has on the sea. Several inches of ice have formed and in the absence of wind, it is in one solid piece across the bay.

Gingerly he tests the ice. It appears firm. First one foot, then the other. Rock solid under his body weight. He does a little jump. What a novelty. Striking out further along the coast, Hurley sees a rare opportunity to photograph the ice cliffs east of Cape Denison from a unique vantage point – out on the sea ice.

His stand camera and its tripod are heavy, but, satisfied that there is no movement at all underfoot, he lumbers further out to sea with his equipment. The cliffs he wants to photograph must be over 50 metres high. They're an awe-inspiring sight, but he'll need to walk a certain distance out in order to fit the majesty of it within the frame.

He sets everything up, splays the legs of his tripod. The need to concentrate initially prevents him from feeling the slight dip of the ice under his feet. Sea water from below is bubbling up through a fissure that has developed, but he doesn't notice that either. Suddenly there's a loud cracking noise.

Hurley's breath is ripped from his chest. The sea engulfs him and drags him down by his waterlogged boots. He cannot yet make a sound, only a gasp for breath. He fumbles wildly for something to grasp to pull himself out on the slippery edge, but the ice is too fragile to hold and it shatters every time he tries to heave his body up. The only thing close enough to cling to is a leg of the tripod, though of course it is too flimsy to use to pull himself out of the water. The current of the outgoing tide is strong; already his legs are being dragged under the ice edge. If he should lose his handhold, his whole body will be sucked under.

A thought seizes him. What of leopard seals? Or orcas? They would gladly take a man and drag him under the surface too. He must get to safety. The icy cold water has already infiltrated every layer of his clothing. His muscles cramp and already his legs have developed the spasmodic movement of a body closing down. His only hope is to reach a thicker wedge of sea ice some five metres distant, where he might clamber up to safety.

Arm over arm, still clutching at the leg of the tripod and sliding it across the ice, his teeth chattering uncontrollably, Hurley drags his frozen body towards salvation, all the while talking to himself to focus his reserves of energy on getting closer to his one chance at survival.

It takes every last ounce of effort to climb out and stagger to his feet, his legs aching and his movements crabbed by cold.

'This is quite the story!' he tells himself. His wet garments stick to his body. A breeze has developed and a thin veneer of ice crystals is starting to form and harden on the fabric. 'Now to pick up the pace.'

The wind is strengthening. Spindrift now fills the air above the cliffs. It will not take much more and the wind will break up the ice in the bay. If that happens and he's still out here standing on it, he could find himself swept out to sea. It's hard to lug the camera with hands so cold they refuse to work and running on frozen limbs is not easy, especially when encased in clothing that is setting like concrete, fast.

Leg over leg, dragging his camera, Hurley lurch-runs across the ice, which bows like a trampoline under such frenzied forward momentum. Only just in time, he reaches the shore. He'd like to collapse onto the snow and catch his breath, but if he lingers too long in these wet clothes he'll freeze to the spot.

Over three kilometres lie between Hurley and the hut. The warm hut, a warming cup of tea. He breaks into a half-run. His garments harden around him until the only movement he can effect is a straight-legged pitching forward, his breath chuffing out before him in a cloud of steam like a locomotive.

When his body appears in the doorway to the hut, very few men notice the frozen man locked into his frozen clothing. The snow that the wind has whipped up and stuck to his outer layers has totally disguised the fact that Hurley is soaked to the skin. He can barely speak, let alone recount his adventure. One by one the other men look up from the mending and the packing of ration bags to see something they have never seen and may never see again – Frank Hurley lost for words.

However, after only an hour or two he has fully recovered and is back to his usual high jinks. Warm in his pyjamas and a heavy overcoat, he dons an old straw hat and entertains the others by singing silly made-up songs while he dries his camera, opening and closing its concertina bellows like a minstrel playing an accordion.

Dr McLean has not strayed far from home but is equally ecstatic. He has spotted a penguin – the first that has been seen for several months. It signals the end of the winter. In the following week, greater numbers return. It is an amusing sight to see them battling the wind in exactly the same fashion as the men have done for the last nine months – heads down, leaning forward on a ridiculous angle. They have reclaimed the rookery that was vacated in the autumn. It's as if they are returning to their favourite holiday spot for the summer, where they will squawk and chatter and bicker with their neighbours about who gets the best position.

Nest construction has started and the earnestness and industry of their works offers a strange echo of the men's building efforts back in January. Balancing pebbles on their feet, the penguins move back and forth, diligently amassing handsome piles that they will sit on as proud parents. The arrival of the penguins may signal the longed-for end of winter but it also signals the arrival of another delight – fresh eggs. Mertz has been talking of omelettes, cooked in butter. The thought gets the others drooling in expectation.

Understandably, healthy appetites motivate the men to make daily rounds of the penguin rookery. Madigan takes Blizzard for a stroll on the end of her chain. The puppy is large, almost fully grown, but remains woefully inexperienced when it comes to the outside world, having spent her first six months locked inside. The dog is intrigued by the penguins and the smells and sounds of the rookery. Madigan leads the dog right through the middle of the nesting birds and the din they raise is astonishing. Bending low, they take aim at the dog like angry, pointed battering rams, shrieking with righteous indignation at the intrusion. Several are brave enough to peck at her hindquarters. Poor Blizzard retreats to between Madigan's legs. She whimpers and pulls her handler back towards the refuge offered by the hut. Yet on the whole the dogs are desperate to get away from the hut and to start ranging more widely. Much like the men themselves. They're itching to explore.

CHAPTER SEVENTEEN

Despite wind conditions remaining horrendous, Mawson has instructed the men to start on sledging preparations. Packing the Paquita ration bags, repairing footwear and adjusting clothing so that it fits better. Long-limbed Ninnis has found that the arms of his Burberry are too short and the wind that whistles through the gap between the top of his mitts and the end of his sleeves pierces his skin painfully when a gale is blowing.

The sledges themselves require modifications as well – decking is mounted to each sledge and plywood packing cases are screwed in place to accommodate cookers and protect sensitive scientific instruments. Even the sledge runners are polished. It takes an inordinate amount of time, but it is crucial.

In the Antarctic environment, to have forgotten vital equipment, even something so small and obvious as tent pegs, is to risk death. Every eventuality must be thought through. But it's a balancing act; whatever they need, or might need in an emergency, represents extra weight. Extra weight requires more energy to haul; more energy requires more food and more food means more weight!

When conditions allow, sledge loads of provisions and kerosene are hauled up to Aladdin's Cave, the supply depot that has been established 10 kilometres away. Having stores cached there might make the difference between life and death. It is entirely possible that foul weather could stall the various sledging parties. Forced to remain in tents for days on end, the men run the risk of eating through their carefully calculated daily rations without making any homeward progress.

Mawson is getting grumpy and increasingly nervous about not being ready in time. If he sees any man idling, he is quick to take him to task. With all their regular work still to do – the scientific observations, dredging, the care of animals, cooking – many of the men are getting impatient with him and feel he's working them too hard. Some believe he treats them like children. All of their stress levels are increasing. Even Ninnis and Hannam start to bicker and have a cursing match over lunch. However, if anything will delay the start of sledging it is the blasted wind.

According to what has been agreed with Captain Davis, the *Aurora* will return on or before 15 January. They must embark on their survey as soon as possible, even if any sane man would consider it an impossibility in such conditions.

Mertz and Ninnis have made huge advances training the dogs in hauling sledges. In the last few months they have experimented with different ways of harnessing them. The Greenland method, which sees the dogs tethered individually to the sledge, is the favoured option. Fanning out, they run in a pack, with one leader taking a slightly more advanced position. To start with, the dogs are unsure of running downwind and are blown about. Ninnis tenses his stomach muscles to brace himself, knowing instinctively that his back will ache from the strain in his bunk tonight.

So far the training runs have been over short distances – ferrying Bage to the transit hut and Hodgie to and from Boat Harbour to read the tide gauge. Three men loaded on the back does not come close to the final weight the dogs will have to pull when the sledges are fully loaded with tents, clothing, provisions, a shotgun, tools, and cooking and surveying equipment, which will amount to around 800 kilograms.

Seals and penguins start to appear on shore in their hundreds. It means the dogs need to be chained up by the hut instead of running free. Throughout winter, Ninnis and Mertz fed the dogs twice a day. Now they receive unlimited feeds. They gorge themselves on cubes of semi-frozen blubber and greedily lick the fatty residue that it leaves on their paws. Afterwards they succumb to torpor - unable to move, their bellies are so full. The dogs grow rounder with each passing day as they build up their own stores of fat. Once they get going on the sledging journeys, body fat will need to power their efforts. Gradually it will burn off, leaving the leanest muscle. The demands of sledging mean that the dogs will be burning more energy than they can take on in the form of food. It is a sad prospect for both Ninnis and Mertz that, once the dogs reach this stage, they will need to be killed. Nothing will go to waste - they will become a source of nourishment to keep the other dogs going.

The men, too, have built up mass during the winter, having had very few opportunities to exercise. Madigan pokes at his belly and thinks of the plum puddings, the fried penguin steaks and the chocolate he has feasted on over the winter months. He has gained six kilos and he supposes it has all concentrated around his middle. He too will need to live off his reserves once surveying commences. Sledging will take a toll on his system; the constant physical exertion of manhauling and the effort his own body will make in keeping itself warm will reduce him in exactly the same way as the dogs. Luckily none of the men intends to eat his companions in a similar fashion.

Four sledging parties will set off in different directions to undertake surveying and scientific work. However, only Mawson's party will enjoy the pulling power of the dogs. It's not that the leader of the expedition is selfish or does not want to exert himself; Mawson's intention is to push much further than the others. He has significant sledging experience and the efficiency that this lends will allow him and his teammates to travel distances that will likely stretch to hundreds of kilometres. Given all their experience in handling and caring for the dogs, Ninnis and Mertz are the natural choice to accompany him on his epic journey.

There is also Bickerton's *Grasshopper* to consider. Bick has been working on the wingless Vickers aircraft throughout the winter in the little hangar they constructed to the side of the main hut. He has fixed runners and brakes to the fuselage where there were wheels. It is hoped that the *Grasshopper* will be able to pull a couple of sledges rigged one behind the other, like a huge propeller-powered draft horse.

Along with the weather, the other major source of anxiety for Mawson is the men's lack of sledging experience. His solution: to send out three reconnaissance parties with instructions to travel close to 100 kilometres, lay supplies and get some much-needed experience of what it means to haul a sledge over uneven ground, to set up a camp and cook in less than perfect conditions, and to take scientific readings and observations while out in the field. For two weeks, the three groups - Madigan, Whetter and Close, Webb, Stillwell and McLean, and Ninnis, Mertz and Murphy - will likely undergo untold hardship. It will be a baptism of fire.

Madigan is terribly excited at the prospect of being first to sight strange new lands, yet he has no concept of how gruelling this experimental excursion will be. At least it will be light for a good twelve hours a day - from 6 a.m. to 6 p.m. - even if they never see the sun through the dense cloud.

Madigan doesn't know what to think. What did he do to deserve Whetter and Close as sledge mates? Is it a compliment or a curse? Does Dux Ipse have such a high opinion of Madigan's leadership skills that he has assigned him the two worst men on the expedition?

It's not that they don't get on. They're nice enough fellows, but Close is useless and Whetter is incurably lazy. It will take great patience on the part of Madigan not to get angry or irritated with either of them.

CHAPTER EIGHTEEN

Madigan has never felt so cold and miserable in his life. He is disgusted with the weather. Dense snow flying horizontally. It took him and Whetter three attempts to pitch the tent in the 100-kilometre winds while poor Close, too debilitated to help, sheltered from the worst of it in the lee of their sledge. The sledge has been picketed with an ice axe. It would come as no surprise if it flew away.

Struggling down into his reindeer sleeping bag, Madigan finds a whole lot of snow has blown into it. What little body heat he has retained after their supper of pemmican will melt it and make him wet, but he can't bring himself to get out into the crystal-flecked air again.

That would be a waste of a hot meal. Besides, his outer Burberry garments have already frozen rigid like an ice carapace across his sleeping bag where he placed them. Instead, he stays where he is, shivering, and stares at the canvas overhead as it tenses and struggles against the tent poles. The daylight through the fabric is sombre, with a greenish tint. How nice to rest the eyes after ten hours of only white.

'Do you think it will hold?' says Close.

Nobody answers. They did not hear him over the sound of flapping canvas and the shriek of the blizzard.

They all fear another loud rip could signal the dramatic demise of what was once a fully functioning windproof tent. Fabric is no match for the mighty southern squalls that have constantly harassed the three men since leaving Cape Denison. The shriek and groan of it is deafening.

After only a few days of sledging, their tent is a tattered thing, held fast with knotted alpine rope. Before Madigan's party left the Cape Denison hut, Mawson and Hurley had modified all the tents. For days they had laboured. This tent has been reinforced in the areas most under strain. But still the wind has won.

Even small rents in the fabric let in a surprising amount of snow – certainly enough to make lighting the Primus virtually impossible. Stitching them up had become a painful part of the daily routine. Mitts off, needle gripped between two stiffened fingers, one stitch, achingly cold hands rapidly shoved into mitts again to warm up. Repeat. One stitch every five minutes. A simple repair takes hours.

A few days ago they witnessed how readily a 130-kilometre wind could transform a small hole into a gash large enough for a man to step through. Needle and thread will no longer be sufficient. Hence the alpine rope. It's a good thing they are on the homeward stretch.

Close's hands are in such a terrible state there is no hope of him helping with anything. With his fingertips blistered and split from the cold, he can no longer do anything for himself – not even secure the ties on his clothing. Madigan has to help him get dressed. Frostbite has affected his feet too.

While the other two men have resigned themselves to completing any task requiring the use of hands, nobody can do Close's walking for him. Neither man wants to carry such a dead weight on the sledge. As long as he can still walk, he'll harness up like the other two and trudge onward. How they long for the safety of the hut!

Madigan's frostbitten eyelid is finally improving, and he can at least open the affected eye, which has been swollen shut for two days. Sight restored offers no great reward – while hauling there is little to see except an endless white vista and one's own trudging feet disappearing in the swirling drift that often reaches knee height.

Whetter's snow-blasted face is peeling, but he hasn't suffered frostbitten extremities like the others. Madigan would say it was due to his incredible laziness – so infrequently does he remove his hands from his mitts to do any work or help with Close. Whetter's fingers seem rather good at snaffling food, mostly chocolate. His apparent greed is starting to irritate Madigan, who at times has worried that their rations may not last the distance. Especially when they were confined to the tent for two days by appalling weather.

Their wind-battered outward journey took them first south, then west. In eight days they covered a staggering 80 kilometres. It is time to turn around. Although the wheel of the sledge-meter that extends beyond the back of the sledge is buckled and broken they know that by the time they reach the hut they will have travelled 160 kilometres. If they get back. As it stands, they cannot leave the tent.

Such distance has been hard won; every kilometre covered feels like five when struggling into the teeth of a gale. Not only have they battled the elements, but also the surface, which has been fiendishly uneven. After the first few days, the slippery blue surface that had been swept clean by the wind gave way to heavily rutted terrain, polished to a high shine in places. The waves of sastrugi have almost reached waist height. Lacking any alternative route, they have had to drag their sledge up and over the top. Had they been travelling over a smooth and flat snowy surface amid sunny and serene conditions they might have covered an even greater distance. They might even have enjoyed themselves.

The one high point is Close's birthday. Before leaving the hut, Mawson gave Madigan extra chocolate and a bottle of port to mark the occasion, but on opening the bottle, they find the alcohol has frozen. Carefully they warm the port over the gas until it can be shaken out into their mugs – a cocktail of lumps and slush. They take swigs, and the uncommon pang of alcohol puts a twisted grin on their faces.

'Happy birthday, Terribus,' Madigan says rather grimly, before making a speech to honour the occasion.

Whetter summons a few words as he helps himself to a bit more chocolate than he is entitled to. Both men then wait for Close's reply.

Eventually it comes: 'I'm too old for this.'

At forty-one years old he probably is. Maybe they all are.

CHAPTER NINETEEN

The menu for 26 September says it all:

Heartiest congratulations on the return of the western reconnaissance party – Madigan, Whetter and Close – after a journey executed in record weather (average wind velocity 58 miles per hour; temperature ranging between minus 17 and minus 37 degrees)

Tension has hardened Mawson's features. His face is careworn and lined. The return of Madigan's western sledging party is a huge relief; had they been any longer he would have sent out a team along their trail to find them. Yet their sudden appearance, with iced-up beards, faces savaged by wind-whipped snow and clothing in a grim state of repair, has not lessened his anxiety. He now knows that the perilous weather conditions they have faced for almost nine months are not peculiar to Cape Denison. The entire region is cursed by foul weather. The entire region is a huge disappointment.

Science is what drives him. Science is now what bothers him. He can see the scientific worth of the expedition diminishing. Spring held such promise, summer even more so. But the conditions are worse than he ever imagined. He had expected a period of calm would accompany the return of the penguins and seals. Instead he is left with the sinking realisation that the dire conditions they faced while sledging prevail. If the weather will not improve – they themselves and their kit will need to.

Mawson thinks back to his previous sledging journeys. He never encountered anything like this in the Ross Sea region, where he was based on Shackleton's *Nimrod* expedition. Fair weather was expected in summer, when cold yet settled days were in plentiful supply. How can they possibly carry out their survey when they are continually punished to within an inch of their lives? Chances are, they will not be able to venture as far afield or do as much fieldwork on the trail as he had hoped.

Despite covering an impressive distance and depositing a considerable amount of food at the 100-kilometre depot, the men have suffered severe physical hardship. The western sledging party was the last group to leave Cape Denison, but all the other sledging parties returned within a matter of days owing to heavy winds. Mawson must assess the men, decide on who should make up the four sledging parties and figure out how he can balance their strengths and weaknesses. He has observed the men all winter. His admiration for some of them has grown; his disdain for others has increased too.

Stillwell tends to sulkiness. He cannot take criticism. Others have moments of inattention and general carelessness, which introduce unnecessary risk into their daily routine. Generally speaking, it is a wonder that they have all got along relatively harmoniously.

Although he can be testy and priggish, Madigan is a good worker and has shown diligence when it comes to his scientific work. His report on the western party's progress was strong. Mawson is indeed impressed that Madigan managed to travel so far with a couple of troublesome fellows – Whetter and Close who have not endeared themselves to the leader at all.

Close's physical condition is particularly bad. Then again, his health has been a source of concern for a while. His rotten teeth in particular. Why the man thought to come to Antarctica is a mystery. And Mawson wonders himself what drove him to select Close.

Whetter is another story. Trouble has been brewing for some time. Among other minor crimes, he was suspected of helping himself to the port while on night watch; yet when confronted, Whetter had denied it. Eventually Mawson caught him out by adding croton oil, which caused a terrible case of diarrhoea. It would be fair to say that Whetter has learned his lesson.

Since returning from sledging, Madigan has expressed irritation at his former sledge-mate's laziness, and wouldn't dream of undertakeing another journey with him. Still, Whetter cares little for what others say. He shamelessly flouts rules and ignores orders. Stern words do little. Even now, when he should be helping dig out the hangar and prepare *Grasshopper*, he strips off his clothes, hops into bed and starts to read a book.

Mawson is furious. 'Whetter, I thought I told you to help Bickerton in the hangar?'

'I've done enough.'

'Enough? There is always, always more to do.'

Whetter says, 'Well, I'm done.'

Mawson is suddenly full of rage. He barks, 'What on earth have you come here for?'

'Not to do that kind of work.'

'You're a bloody fool to come on an expedition if that is the case.'

'Bloody fool yourself. I won't be caught on another one.'

Mawson can scarcely believe his ears. His disrespectful tone, his appalling attitude! All expressed in the presence of others in the hut. Such disdain for a leader's authority will not be tolerated. Mawson takes a moment, draws breath then calmly asks Whetter to step into his room for a private word.

What passes between them can still be heard quite clearly; the walls of Mawson's room are not thick and do not go all the way up to the ceiling. However there is a semblance of privacy, further enhanced by the hushed tones of their conversation.

'We never get a moment's peace,' Whetter complains.

'Your official duties during winter were minor compared with the workload of others. You merely had to bring in twelve boxes of ice for hut use, store it in the verandah. While you were out sledging, Laseron took over. He had it all finished within an hour.'

Whetter scoffs. 'Can you blame me for dragging it out? You're always finding work for anyone who doesn't appear busy.'

'Yes, indeed I am!'

'So we make sure to look busy, even when we're not.'

'You don't think I haven't noticed? All winter. You and the other two – you've drawn out any task so as to not get saddled with additional work. It's deplorable.' Mawson feels his temper rising.

'I want to knock off at four o'clock. If we start at nine and have a one-hour break for lunch, that's six hours.' He pauses. 'Surely that's fair.'

Mawson does not like the manner in which Whetter has gone about it, but his suggestion is reasonable. 'Whatever arrangement is reached, everyone must be treated equally. I will consider it and let you know my decision at dinner.'

'The united efforts of all are required to make this expedition successful. As leader, I am sure to get a certain kudos, but our success is for all, not just me.'

There is a general hum of agreement around the table. Everyone has been whispering about the stern words that passed between Mawson and Whetter that afternoon, but they are not expecting what comes next.

'I've decided to institute a new routine. Work will be carried out from 9 a.m. until 4 p.m. with a one-hour break for lunch. Sundays will be a day of leisure when you may please yourselves.'

A chorus of eager voices, undoubtedly led by Hurley, rises in spirited yahooing.

Mawson smiles but says loudly and firmly, 'Of course, on fine days when outside work can be accomplished, all hands are required to do anything I ask them to do. There is still so much to do in order to get our sledging program underway in just under a month.'

'Somebody tell the wind it can have a day or two off,' comes an anonymous shout. 'Surely its work here is done.'

PAQUITA

1909

The Prof doesn't mince his words: 'Mawson was the real leader and the soul of our expedition to the magnetic pole,' he tells the assembled dignitaries at the official reception. 'A man of infinite resource, splendid physique, astonishing indifference to frost.'

Arriving in Adelaide, Mawson is greeted by a hundred cheering students who rush to his carriage and lift him high onto their shoulders. Much to Mawson's amused embarrassment, they carry him all the way from the railway station to the university.

It's at the university that she notices him. He looks unassuming, modest, standing there in his sports clothes on the wet grass, in the mist. Of course, everyone knows him by reputation. His polar adventures with Sir Ernest Shackleton and Professor Edgeworth David have made him something of a local hero.

How interesting to see him finally – the famous Douglas Mawson. Perhaps others would notice his height first – he is a tall figure. Other girls might notice that he is good looking, slim and athletic. But it is the grin that strikes Paquita as more beguiling. And the way he turns when a friend says something to him and offers such a warm reply. She can tell instantly that he is an authentic person. He may be famous but he's unaffected by his fame. That much she can tell from twenty paces.

It is only a matter of weeks and the twenty-seven-year-old Mawson appears at dinner in Broken Hill, where Paquita and her sister are staying with their father. He went overseas after that, on a tour. But later, on his return, there will be many dinners and weekends at her family's summer house on the coast. That is where they fall in love, looking at the sea, with the sound of waves, and no ice in sight.

'Will you marry me?'

Paquita's answer is plain to see.

Yet, before any of that can be settled, there's an expedition to organise and a great scramble to find money. There's also an entire year in Antarctica to endure and possibly a book to write on his return – he's been offered a thousand pounds to write about his experiences. It's all part of the exploring game. In order to write the book, he'll need to do some serious exploring.

But Douglas is confident of success. In both expedition terms and in affairs of the heart.

CHAPTER TWENTY

Black Sunday, Hannam calls it. The thirteenth of October. A Sunday like many others, except a particularly violent blizzard is raging – a real howler. All day the wind has been throwing the most dreadful temper tantrum, punctuated by sudden, terrifically strong gusts that blast the hut like a detonation. In its rioting fury, the wind upturns boxes full of stores from the windbreaker, flings gravel and loose rocks, and scatters huge blocks of ice. Not satisfied with the havoc it has already caused, the wind picks up and tosses a sledge end over end to the brink of the sea.

Their shelter groans and shudders in response. The clothing hanging from the rafters swings lazily back and forth as if in a breeze. Whenever a particularly loud creak or crunch is heard above their heads, faces tip back and eyes dart about the ceiling with morbid interest, but nobody is terribly concerned.

Madigan continues with the dinner preparation; rolling fresh penguin breasts in flour and onion powder for frying. Defrosting three frozen cans of plum pudding inside the oven and hoping they don't explode.

Later that evening, over dinner, there's an indistinct *boom-crack* from outside, then an almighty crash. The hut shudders in response. The men exchange frightened looks. A few leap to their feet, terrified that the roof might be about to lift off.

'What the hell was that?'

'The roof is breaking up!' Terribus cries.

'No, it sounded further away.'

Alfie, one of the first men on his feet, starts to get kitted up. As chief architect of their hut, he feels responsible for its structural integrity. Mawson, too, drags on his Burberry.

As soon as Alfie sticks his head out of the trap door he sees.

'The radio mast!' he shouts down to the others over the din of the storm.

Mawson appears by his side and curses. The mast has been carried away by the tempest, and so are his words. He ducks his head low into the shelter and out of the wind.

'Did you see?' Alfie asks. 'The broken length of it lying across the terrain just outside the hut.'

Mawson had indeed seen the tangle of splintered wood and wire that is all that remains of their proud achievement, their great hope for communications. Not that they have had much luck with it. Who knows if the messages that Hannam regularly taps out travel any great distance. If they're picked up by Macquarie, there's been no indication – only static in reply.

'Blimey! Just think, if any of us were in the way . . . we'd have been mincemeat.'

'We'd be mincemeat anyway if we were out in that hurricane.'

'The south mast is looking a bit dicey,' Mawson says. 'That one would fall on the hut. One of those gusts could easily take it out.'

Either brave or foolhardy, Mawson ventures from the trap door to go and inspect the puff-anemometer that he set up in that morning. He keeps his body close to the ground like a soldier advancing on an enemy. Whenever a strong gust hits, he drops, lying flat to avoid being picked up and thrown about like a ragdoll.

Of course, the Puff-Annie has been demolished, fallen victim once again to the blasting winds of Cape Denison, but not before recording wind gusts that had reached 325 kilometres an hour. Nobody would believe the reading, had they not heard the roar of the wind and felt the buffeting force of it from inside the still partially buried hut. In fact, given the reading is at the absolute maximum the instrument can record, Mawson suspects that the gusts might have been even stronger.

Inside, word soon spreads about the possibility of the southern wireless mast falling on the hut. Bick and Ninnis in the top bunks feel particularly vulnerable and gather up their blankets to doss on the floor. So do the men who sleep in the lower bunks on the eastern side of the hut. The only sensible thing to do is dismantle the southern mast. But there will be no opportunity to do that until the weather has calmed down enough. For now, it's better to take precautions and opt for alternative sleeping arrangements.

They all lie awake and listen intently for any indication that total annihilation is imminent. As uncomfortable as it is, they cannot help but feel grateful. They could have been caught out in this while sledging. Perhaps they soon will be.

CHAPTER TWENTY-ONE

It's the beginning of November and at the first sign of fine weather, they'll be away – four teams of three men and a fifth team with the *Grasshopper*. More than a few of the men have started to pack up their personal belongings. The day they have all circled in their diaries is 15 January. This is the day the ship must leave, so they will all have to be back from their sledging journeys or they'll miss their only way home. Some will be away for six weeks, others longer. Everyone needs to pack up their things before they set out, in case the ship returns ahead of schedule. Some suggest the *Aurora* could arrive by Christmas.

The reunion with Captain Davis will be a glorious one;

the ship carries letters from loved ones and newspapers, telling of what has happened in the world since they were deprived of all contact. They'll all tease good old Gloomy who, like last year, will probably complain that the coal is running low or that the season is too advanced or the crossing too perilous. He and Terribus would make good companions.

Now that their time at Cape Denison will soon draw to a close, a certain ambivalence takes hold in their hearts. How strange that this inhospitable place has become so familiar as to promote a sort of nostalgia - the bay with its bergs and snow-covered islets, the Antarctic plateau sharply rising behind them, the rocky spined ridges with their bursting boisterous populations of penguins flanking their hut. These are the real, concrete elements of their existence. By contrast, Australia seems like an imaginary world - streets, women, ordinary clothing, trees - these things are quite fantastical.

Madigan is apprehensive about what lies ahead. He fears it could be as bad as their spring excursion, and that was brutal punishment. They will be gone for much longer this time; he will need to be more cautious, and diligent about protecting his hands and feet from frostbite. He prays their newly reinforced tent will stand up to conditions better than it did during their reconnaissance journey. There will be nobody to call on for help should they need it.

Out there, aside from one's sledge mates, one is alone. Whenever there are delays caused by foul weather, anxiety redoubles about whether rations will be sufficient. Then there are the disturbing thoughts of crevasses, or injuries preventing men from walking, pulling.

Despite his misgivings, he is enormously proud of what he was able to achieve; he intends to distinguish himself. If he's taken a year out from his university career, he's going to make something of it. Madigan is grateful to be paired with Doc McLean and Correll. He won't have greedy guts Whetter to contend with. Nor will he have to dress and undress and manhandle poor old, broken down Terribus into his frozen finneskoes. With the Doc and Correll, he's sure they will make excellent progress eastward along the coast.

Fond farewells are the order of the day, with the four friends at Hyde Park Corner soon going their separate ways – Bick, Madigan, Mertz and Ninnis have stripped bare the walls of their bunks, folded away all their clothes and letters and photos and stacked their books into packing cases. Personal belongings are strictly limited to the bare minimum for sledging. No excess weight will be tolerated when manhauling.

Madigan says, 'Good luck with the *Grasshopper.* I will sure miss you, Bick. But probably not you two,' he makes a comical face at Ninnis and Mertz. 'But disappearing with

Dux Ipse and all the dogs as you are, you scarcely need me to wish you luck.'

'I will miss you, Madi,' says Ninnis. 'It doesn't matter how rude you are.'

'Or how bad your German be,' adds Mertz.

'How bad your German *IS*!' shouts Madigan.

Several weeks ago, Ninnis and Madigan engaged in a quiet heart to heart about their sledging experiences. Madigan makes no secret of his opinion – the sheer awfulness of being so cold, hungry and thirsty. But pride is a funny thing, and amid cheering thoughts of distinguishing himself again, he has forgotten the worst of it.

Ninnis has not admitted to anyone else, or even acknowledged it to himself – yet he is totally in agreement about the sledging. He feels rather ashamed of thinking this – after all, it is what he wanted above all else to do with his life – be a polar explorer. Perhaps it is different in calm weather. But in a gale and freezing temperatures, there can be nothing worse.

Packing for home he has felt ecstatic. He has gathered up some colourful geological finds from the moraine to take home as souvenirs as well as presents for family. Will this be his last time in the south, he wonders. He's not so keen on returning. He writes his name on the tin-lined crate that he has painstakingly sealed up. He cannot wait until the time he can unpack it.

Six members of the Australasian Antarctic Expedition in 1911. Back Row: Percy Correll, Cecil Madigan and Frank Bickerton. Front row: Alfred Hodgeman, Sir Douglas Mawson and Morton Moyes, meteorologist from Frank Wild's Western Base party.

Penguins investigate the *Aurora* anchored to floe-ice off the Western Base.

After the worst of the weather front has passed, the unloading of the ice-sheathed *Aurora* can resume, but the blizzard has dispelled all romantic notions of Cape Denison being anything other than a bleak and exposed outpost.

National Library of Australia / Frank Hurley

The two huts at Cape Denison were surprisingly solid despite being erected by men with no building experience.

State Library of New South Wales / Frank Hurley

Lunch prep gets underway in the main room while those not on mess duty continue their activities in the workshop.

National Library of Australia / Frank Hurley

Collecting ice for melting down over the stove is the worst job. Heading outside in all weathers with an icepick, chipping away while battling the wind.

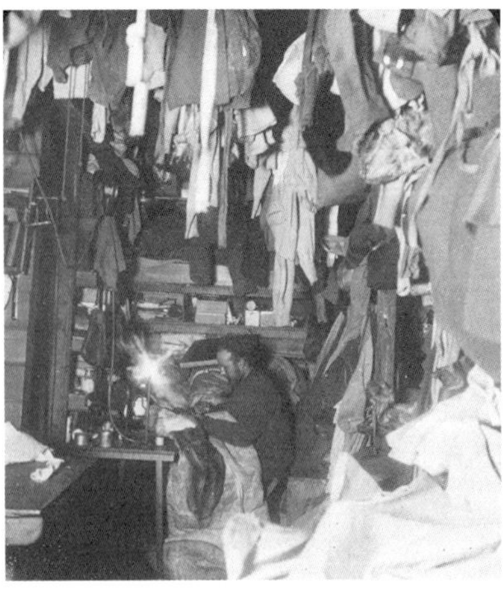

National Library of Australia / Frank Hurley

Bage mends his sleeping bag amid the bunks arranged in two tiers and a veritable forest of outdoor clothing drying from the ceiling.

State Library of New South Wales

For weeks Murphy had been digging out the stores and the covered verandah each day. Every day it would fill again with windblown snow so that the only way in or out was to slide in on one's belly. It was an endless task, a thankless task, a hopeless task. Murphy versus the Antarctic winter.

State Library of New South Wales

Ninnis and the dogs taking a heavy load of stores to Aladdin's Cave.

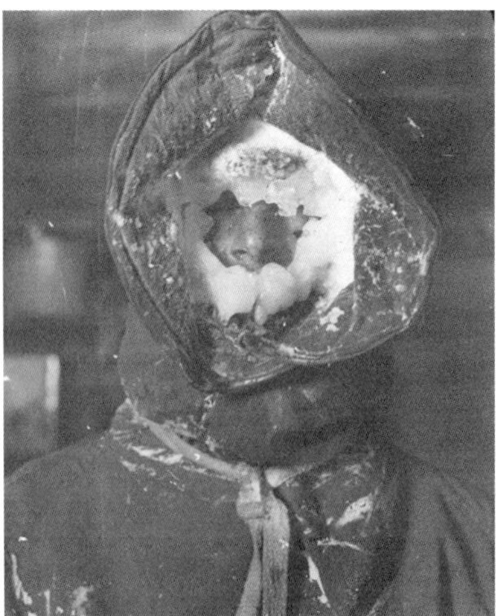

National Library of Australia / Frank Hurley

The men's own laboured breath freezes into their beards, encasing their faces within their Burberry helmets behind wreaths of delicate icicles. Each man feels shut off from the world.

State Library of New South Wales / Frank Hurley

Once the two huts are completely snowed under, the only way to leave is by the trapdoor on the verandah roof.

State Library of New South Wales / Frank Hurley

With a few well thought-out adjustments, the wingless plane, which has been christened the *Grasshopper*, will be used to haul sledges as a strange sort of hybrid air tractor.

National Library of Australia / Frank Hurley

Sastrugi sculpted by the incessant winds challenge the men and dogs – often their impression is of crossing an inland sea bleached of all colour.

State Library of New South Wales / Frank Hurley

It occurs to Mawson that the heavy sledge could be cut down successfully – the difference between life and death.

However, if there is one good thing about braving the elements again, it is the team he will be joining. Ninnis is delighted to be paired with Mertz and Mawson. As they expect to cover the greatest distance, they will take all the dogs. The far-eastern party will venture hundreds of kilometres and hopefully their survey will meet up with the survey conducted by those of Captain Scott's men who have ventured west from the Ross Sea while Scott makes his bid for the South Pole. Perhaps his name will end up on the map.

When the day of departure finally dawns, Ninnis writes in his diary:

'We and the Southern Party leave tomorrow. I must close my writing now, maybe for two months, maybe for good and all, for who knows what may happen during the next two months. The hut seems very forlorn, with only nine fellows left. I hope we shall pull through all right, and join up with the 'Terra Nova' farthest; and now I will stop.'

TOWN HALL ADELAIDE

FAREWELLS

The thunderous applause carries on for some time. The Governor was the last of this evening's speakers, all of them praising the forthcoming Australasian Antarctic Expedition and its ambitious leader. Mawson gets to his feet and waits for quiet. The cheering from the audience continues. He surveys the room with a slightly puzzled expression. While he certainly hopes to bring glory to Australia, he cannot take all the credit, it will be a group effort. He clears his throat.

> *'Most of the speakers have referred to myself as the chief spirit in this expedition. To some extent this is so. But I am only one*

of the land party. I can do nothing more, in some respects, towards the success of the expedition after having chosen the staff. The most important point to look for in members of an expedition like this is character.

I think it can be confidently said that the men representing South Australia are well fitted in that respect. It is impossible for me to tell how they are going to act until circumstances arise that will bring out character.

In the land of desolation, in that land of loneliness to which we are about to proceed, there are conditions to measure a man at his worth. You will know in eighteen months' time who have been the successful members of this expedition and who have not. I personally feel that all the men chosen will be successful.'

The air fills with clapping and cheering. This time the audience gets to its feet. Mawson lets the applause wash over him. He is the youngest commander to ever venture into the Antarctic and this is the youngest group of men to be led. Mawson hopes he can also lead them back to safety.

CHAPTER TWENTY-TWO

The first to leave Cape Denison are the southern support party, the men tasked with hauling extra supplies for Hurley, Bage and Webb's southern team, who will head for the magnetic south pole. After all the waiting around, it is an immense relief to be finally away; even if it is in fits and starts, rather than a grand departure with the teams fanning out in different directions, into the unknown. The weather is proving good enough, but not great.

The sledge has been fitted with a bamboo mast and spars on which a sail can be rigged to aid them in hauling. It is quite a load. There's the tent and sleeping bags, extra clothing, rope, pick and shovel, skis, crampons, theodolite legs, sledging rations, dog food, seal meat, blubber and pemmican, a cooker, navigational equipment, the binoculars, the hypsometer, thermometers, photographic gear and medical kit. There is also kerosene, carried well away from food in screw-top tins. Fuel is as precious as the food; without it they wouldn't be able to cook or melt snow for drinking. The last meal they eat together on the eve of departure will live on in memory. Nothing they will be eating for the next few weeks will taste as good as the fried penguin eggs and bacon, the plum tart and chocolate.

Having already put off their departure by a couple of days due to bad drift, Murphy, Hunter and Laseron scarcely hesitate when the blizzard finally subsides enough for them to strike out. With the utmost bravery they confront the bracing wind, their outlines disappearing into the white haze. Yet it comes as no surprise when they return to the hut later that afternoon, having abandoned their best intentions and their sledge several kilometres inland. The fearfully strong gusts up on the plateau make further progress an impossibility. No point hauling the heavily laden sledge back down from the plateau. Happy to give it a few days, they will return when it's calmer to dig out the sledge and get underway. They hope!

The near east coastal party of Madigan, McLean and Correll, aided by Bickerton, are next to head away up the steep ice ramp to the plateau; then Stillwell, Alfie and Close, who will support them, hauling supplies as far as they can before returning to the hut where Murphy, Hodgeman and Hannam remain behind. The last to leave are the dog party of Mawson, Mertz and Ninnis. They'll outstrip the others with the dogs being so fast, even with three sledges each weighing over 250 kilograms.

It is a reassuring notion that the sledges will get lighter as they eat their way through eight weeks of provisions. In the case of Mawson's party, they also have the added weight of food for the dogs. Ninnis and Mertz have spent weeks drying seal meat and blubber over the stove to reduce the weight. It won't be quite enough for the dogs. The weaker dogs themselves will need to be a food source for the strongest pullers as the weeks go by and their energy reserves dwindle. For the two men who have devoted themselves to the care of the animals for so long, it will be sad to have to destroy individual dogs. But they have come to accept it as a necessary part of exploration.

Despite the foul weather, jitters and abortive starts, their spirits are high. After all, this is the expedition's *raison d'être*. To discover new lands. The four parties will rendezvous at Aladdin's Cave, where they can take on provisions, before re-lashing everything to accommodate the extra food and kerosene supplies that were cached there over the last few months. Then they will spend one last night together in the relative luxury of the ice cave with its abundant food and sparkling ice walls before venturing to the east and south, forging into the unknown, their own separate ways. Their sledge-meters will record the extent of their travels. Around and around, the bicycle wheel will spin, accumulating distance while the men accumulate experience – both hard won and unforgettable.

CHAPTER TWENTY-THREE

Navigation is never easy in the absence of notable landmark features. And straying this close to the magnetic south pole, the compass does not tell the truth. Mawson has also discovered the compass housing is not airtight, and the wind blows the needle about. It is most irritating. Another frustration is that they often lack a discernible horizon. Sometimes they lack the conviction that there is even a sun in the sky, so diffuse is its glow.

Mawson slips the page he has ripped from the Nautical Almanac back into his sledging notebook while Ninnis places the theodolite back in its box. A single set of readings for azimuth is all they have managed. It is too cold to do more. They usually take angles to establish their position at afternoon tea when they can wrap numb hands around a pannikin of hot cocoa while sheltering from the wind in the lee of the sledge. They are constantly risking frostbite. And yet everything seems to require un-gloved hands. And there are calculations to do in pencil after referring to tables, while keeping track of the magnetic variation between true south and the magnetic south pole. Cold hands never work efficiently. Neither does a cold brain. There are complications, errors, adjustments.

Mertz is in charge of the hypsometer – the higher their elevation, the lower the point at which the water in the instrument boils. It is a far more straightforward task. Wind and weather conditions are carefully noted several times a day so a comparison can be made with details recorded by the other parties. Their readings and records anchor them to reality even when the real world feels fluid and unstable. So easily one's senses can be whipped away.

The bright light that engulfs them does not cast any shadows; nor does it reveal obstacles in the form of recesses and depressions. It is so very hard to see the surface when everything is bled of definition – there are frequent stumbles and their gait assumes a tripping and lurching character. It would be funny, were it not so jarring. Yet they are making progress – sometimes 20 kilometres a day, often more.

And so they continue across the coastal highlands towards a vague and blurry zone – east is their destination, and there is always more ground to cover. They could continue eastward and eventually they would end up where they started.

Somewhere to the north lies the sea. Sometimes it is sheathed in ice or lashed to whiteness by the wind, so although it is there, it is seldom visible. Somewhere far behind, Madigan's team hugs the coast, their pace limited by the need to haul their sledges, sometimes across land, sometimes across the frozen sea itself. Ninnis wonders if time passes for them as oddly as it does for him.

The route of the far eastern party is by no means flat. Upwards they climb, topping out on the rounded backs of ancient accumulations of snow and ice, urging the two dog teams to maintain a strong stride, a steady pace. Sharp descents lead them inadvertently into depressions laced with crevasses. The dogs are brought up short. Their noses search the air for a reason for the sudden halt. They understand little of this life of toil. Sorrowful howling is their response whenever a snow bridge collapses and they fall suddenly into the void. Always to the length of the harness, and so they are easily retrieved. From the chasm they emerge, their wild-eyed scrambling communicating nothing but utter confusion.

The breath of the dogs freezes on their ears, topping them with fine rime which comes to resemble delicate fretwork. It gathers along the length of their harnesses, hardening the leather until it is as stiff as bone and the sledge's traces resemble an external skeleton of some large white beast with thirty-two legs.

The men's own laboured exhalations freeze into their beards, encasing their faces within their Burberry helmets behind wreaths of delicate icicles. Each man feels shut off from the world. No conversation passes between them. Memories of the camaraderie of hut life seep away until all that exists is the present, the days of walking, of riding the sledge when conditions allow, taking observations, noting the meteorological conditions, mapping their route, pitching the tent, feeding the dogs, and wolfing down their simple meals. And the wind. Always the iron wind.

CHAPTER TWENTY-FOUR

Supper looks disgusting. It might get Ninnis elected to the Crook Cooks Association had they been in the hut. He is thankful that cold and hungry men appreciate anything hot – especially a sloppy brown stew, hastily rustled up from a brick of pemmican, some crumbled sledging biscuits and melted snow. Gulping it down before it has a chance to cool, they welcome its restorative warmth into their bellies. None of them can imagine anything more richly satisfying. The only thing that would rouse their spirits further would be the prospect of a second helping of brown sloppy stew.

'This hoosh is delicious,' says Ninnis cheerfully. 'I've outdone myself.'

'Delicious,' Mertz agrees. 'You have.'

Mawson makes no comment. He has other things on his mind. 'Slower progress today. Those seracs rising up from the Drygalski Ice Tongue posed a frightful challenge. Enormous folds.'

They lost much time going down the steep valleys of the pressure ridges, many of them half a kilometre wide with crevasses choked up with loose snow. The effort expended by the dogs was considerable but ultimately fruitless. Several times they hauled up the other side only to encounter some impassable obstacle.

'Such a waste of energy pressing through only to find yourself in a dead end.'

Murmurs of agreement sound amid eating and the insistent flapping of tent fabric.

Mawson raises his voice, 'How did the dogs go? With Jappy and the pups?' He glances up at Mertz over his steaming pannikin.

Mertz frowns as though the question barely warrants a response. 'They love it. Of course. It is meat.' He pauses. 'Not Shackleton. He wouldn't touch it. But the others were happy to eat Jappy.'

Ninnis grimaces at the thought of the unfortunate pups whose only crime was to be born on the trail. They had to be destroyed along with Jappy. The mere mention of the deed sours his stomach. He left the butchering of the poor spent creature to Mertz. He will have to take a turn himself in due course.

'How many feeds will we get out of her?'

Mertz slurps down the last of his hoosh, wipes his mouth and says, 'What was that? I didn't hear you.'

Mawson clears his throat and asks in a loud clear voice, 'How many feeds will we get out of her?'

'Another one, definitely. Maybe two.'

'They're ravenous,' says Mawson. 'They're working hard. A few look close to exhaustion. Tomorrow I want you to harness all the best dogs to the second sledge – with the food. We'll send the weaker ones first with a lighter load.'

Mertz nods.

'And your shoulder, Ninnis? You had a narrow escape this morning.'

'Don't I know it.'

Directly on leaving the tent at the last camp, Ninnis broke through a snow bridge. The crevasse on which they had inadvertently pitched their tent was a deep one. It could have swallowed them whole had it collapsed while they slept. In the event, only Mawson's camera suffered. The strap snapped, caught on Ninnis's arm, which he flung out to save himself from falling to his death.

'My shoulder is a bit stiff but it will not disrupt us from our purpose.'

Mawson nods but doesn't look up. His mouth twitches as he remembers the Prof, how silently he was swallowed up. How politely he called out to raise the alarm. He has no doubt Mertz and Ninnis would yell like hell.

Their meal is over. Without another word, the three men wriggle down into their sleeping bags and disappear into their own private thoughts, which lead swiftly into sleep.

CHAPTER TWENTY-FIVE

For weeks they forge a path across a blank landscape that spreads infinitely outwards, three dark figures, a procession of dogs and sledges spoiling the pristine surface in the name of science with their plodding footsteps, yellow patches of snow and dog droppings, and channels left by the sledge runners. In calm weather Ninnis plays his mouth organ. Sometimes Mertz yodels.

The hard ice slopes of the eastern coastal highlands are far behind them now. They have battled acres of crumpled pressure ridges and picked a way through depressions riddled with crevasses. Pillowy undulations fill their field of vision for several days before the wave-like sastrugi return, giving them the impression that they are crossing an inland sea bleached of all colour.

There are the uphill slogs, and the sharp descents when the sledges threaten to overrun the dogs or capsize completely. Sometimes calamity strikes and they have to unhitch the dogs and reload the sledge. In the mornings, when they stop for chocolate, they breathe deeply. Perhaps they admire a shaft of sunlight or a bank of dark cumulus cloud massing on the horizon. It certainly takes the breath away when the sky opens and reveals tiny patches of blue that become big and square, and they turn their faces to the sun and rejoice in its warmth.

A different sort of excitement takes hold whenever an unusual sight presents itself and breaks the monotony: a volcanic crater, a large rock, a *nunatak*, a mysterious icy opening in the surface littered with so many fine stones, it could be a quarry. One day, out of an empty sky and far from the coast, a snow petrel flies into Ninnis's sledge.

'Hey! Look! Look!' he calls to his companions. The delight and surprise in his voice is infectious as the bird flies off into a cirrus-streaked sky to goodness knows where.

Sometimes they need to rope up the sledges for safety and together with the dogs, Mawson and Ninnis haul their cumbersome loads through the worst of the tricky terrain while Mertz presses ahead on skis to scout the way through danger. In his strong clear voice he sings his student songs.

In early December they get glimpses of coastline and whale-backed islands which they duly note on their map; they won't pass this way on the return journey to the hut.

There are days when the dogs become completely encased in snow. Sometimes the men's clothing freezes solid on their bodies. Sometimes winds threaten to demolish their tent. The fabric convulses under the assault, protesting with mad flapping. The sound floods their ears so that they must shout to be heard. It roars as though their camp is an affront to the vast emptiness these three men have trespassed upon. On such nights, Mawson insists they take the most precious belongings into their sleeping bags as a precaution – all the things that could blow away should their tent be ripped asunder.

Overall, the dramas are short-lived and of little consequence. One of the dogs rips a hole in a ration bag and eats all the butter. Another gives birth to more unfortunate puppies and these must be dealt with as humanely as possible. Snow blindness is a regular complaint. Mawson develops a migraine. Ninnis's finger is infected and needs to be lanced.

Their fitness is improving. They take turns as leading man of their little procession. The frontrunner sets out at a distance and then maintains a jogging pace that the dogs find irresistible to follow. Further and further they venture from Cape Denison – they have travelled more than 400 kilometres across the continent's white face.

Another demon of a day is ticked off. How the character of the surface changes! Hour by hour, day by day. Now it is dead flat and crusted; now pillowy and smooth. When the weather is calm, the snow squeaks. It might be old snow that crumbles like biscuit, or with a hard shell so it may be picked up in a slab. In warmer weather it is as heavy as cement from a bricklayer's trowel; or soupy, almost viscous as oil paint applied with a thick artist's brush.

Their days are filled with kerosene and tent poles and savoury smells of pemmican stirred over the Primus, cooked and consumed within ten minutes. There are snacks of chocolate throughout the day

There are periods when the sun strikes and the surface sparkles like a cloth woven from diamonds. Sometimes the vista is so beautiful that a cheerful sigh will escape from Ninnis's lips, or Mertz will yodel in appreciation. They are tired. They are content. Nothing prepares them for what comes next.

It has just passed noon on 14 December. When the ground collapses under Ninnis, he falls sideways; after him slide the sledge and dogs. Together they tumble, a jolting disorder of crunching wood and bone, the whip-snap of harnesses and tent poles and metal runners. Boxes spray their contents, the tent bounces and balloons. Surprise is still caught in his throat on impact.

CHAPTER TWENTY-SIX

They are frantic. It is unthinkable. It is an unreal thing, this void in front of their eyes, this gaping hole. And Ninnis gone; swallowed whole. How is it possible? Without a sound, leaving no trace. Hastening to the edge of the crevasse, Mawson and Mertz call his name, their stunned expressions barely registering their friend's hideous fate.

'Ninnis!' Their voices echo downwards. In desperation they call it again and again. They stare dumbfounded into the depths. 'Ninnis!'

All is silence.

Ninnis could not have survived such a fall. The sides of the crevasse are sheer; the drop precipitous. They cannot see much. Certainly not the bottom. There is no trace of the mangled sledge. No body. The lengths of alpine rope they tie together barely reach halfway to the ledge far below, where one dog is splayed awkwardly. Hanging over the edge and peering into the gloom, they can just make out a lighter shape that looks like it might be one of the canvas ration bags. Like the miserable whimpering dog with the broken back, it is unreachable.

For several hours they linger by the hole in the snow, listening for any indication that Ninnis is still alive, calling intermittently. Thankfully the sky is clear and winds are light. The remaining dogs grow restless. They nip at each other in irritation and confusion at the unexpected halt, while a few drop to the snow in an attitude of watchful repose. Little thought is given as to how they would mount a rescue should they hear any sounds of life drifting up from below. The end must surely have come. It has for the injured dog, which is now silent. Yet something anchors them to the spot. Is it indecision? What can they do but shout once more: 'Ninnis!'

Nudging their thoughts from Ninnis, they try to figure out what happened. Both had crossed the same area before him. Why had neither Mertz on his skis nor Mawson's sledge triggered the collapse?

'My skis spread the weight.'

Mawson 'I was sitting on the sledge as I crossed over. I thought it looked suspicious. That's why I took it at an angle. I called to Ninnis to watch it . . .' Mawson pauses, thinking back to the moments before the tragedy. 'Ninnis was walking beside his sledge.'

Mertz hangs his head in despair.

There is nothing more to say. They must get on. Mawson climbs a nearby hill and takes sights. Their position is 68 degrees south, 151 degrees east. Returning to the crevasse, they call one last time. There is no response. It is late evening. They have been there nine hours and must get going.

'Right, we should read the burial service,' says Mawson, steeling himself to the task.

Mertz begins to weep. He cannot bear to think it has come to this.

They have lost their friend. But they have also lost half of everything. Together they take stock. The situation is even worse than imagined. The second sledge was packed with essentials, precisely because it was assumed that any weakened snow bridge would be triggered by the first sledge to go over it. For that reason, all the best dogs and the essential items were loaded onto the second sledge.

Ninnis's sledge carried the tent, the tent poles, the ground sheet; three and a half weeks of provisions for three men, and all the dog food. They've also lost the tools, Mertz's windproof overtrousers and his Burberry hood; their cups and spoons are gone, as are the mast and sail to harness the wind. And they are 500 kilometres from the hut. Looking over what they do have, they count the Primus, the cooker and fuel; the theodolite, the tent cover and their sleeping bags and private kit, and the rifle. While it may not be cause for celebration, it is a source of relief. They can melt snow for water and heat food, they have means of getting warm and a way of navigating homeward. In many ways they are better off than they would have been, had their sledge been the one to go.

'We'll have to reduce our rations to make our supplies last longer.'

'The dogs have nothing,' Mertz says.

'Well, we will need to eat them.'

'It's good we have the rifle.' Mertz pauses. 'And the cooker. Can't eat raw.'

Mawson says, 'Shelter. That's our first priority. We must have shelter.'

'We have the spare tent cover,' says Mertz. 'But no way of putting it up.'

There's a clear advantage to backtracking to their last camp. The third sledge was left there. They could use the runners for tent poles. There's little beyond that to salvage, but it's worth checking. They have definitely not left any supplies. In fact, they have not laid any depots or cached food or equipment, because they never anticipated covering the same terrain on the way back. They are certainly in strife.

'We've got nowhere to sleep.'

'I cannot sleep,' says Mertz.

He is right. Even if they had a place to lay their heads, sleep would not be forthcoming. Anxious and keyed up by the scale of their challenge, they will set off immediately. Retracing their steps, they have a 20-kilometre hike ahead of them to sort out some shelter. One problem solved. They will get there on adrenalin. But the dogs? The dogs are already hungry.

'May God help us,' says Mawson.

CHAPTER TWENTY-SEVEN

They arrive at their former campsite at 2.30 in the morning. They have covered almost 50 kilometres in one long push through the night.

They cobble together a shelter using the spare tent cover. Whatever might pass for a tent pole is put into service – a sledge runner and a ski. The result, a sagging and forlorn shapeless thing, scarcely resembles their former taut and efficient accommodations. It looks like a skeleton with skin hanging off it. Too exhausted to do anything else, they fall into a deep slumber.

By ten the next day they're awake and confronting their new reality. Their new tent's sides are even more sunken now, collapsing under the weight of the snow that fell in the night. The weatherproof japara fabric is far from breathable. After sleeping in the awkwardly pitched, crumpled space with its reduced airflow, the men wake to see their breath has collected then frozen on the underside of the fabric. Iced up, the tent, will be far heavier than it was.

George is the first dog to be slaughtered, chopped up and fed to the other five dogs, who seem insensible with hunger after hauling such a distance on empty stomachs. Mawson cooks some dog meat for their own breakfast. It crackles on impact with the hot metal of the billy can and emits strange fumes, meaty yet unfamiliar. Neither of them find the meat at all palatable, but it is something they will need to get used to. Also eating off wooden spoons fashioned from bits of sledge.

Discussing their predicament takes all their mental energy. They have so little food. They need to travel as quickly as they can. That means reducing the weight of their sledge so the load on the dogs is not so heavy. They are ruthless, jettisoning anything that is surplus to their immediate requirements.

'I never thought I would do this,' Mertz says, throwing the reindeer mitts and excess leather straps to the dogs to eat. In the past they would have been given the whip for such transgressions.

It's a warm day with no wind. They discuss the possibility of leaving straight away, but decide the snow is too soft to make any reasonable headway. It is better to travel through the night when it is still light, but temperatures are lower and the surface firmer. The route they'll follow is still up for debate.

Mawson says, 'We're taking a risk traversing this dangerous stretch of the plateau.'

'What if we travel down to coast? We can eat seals and penguins. Travel over sea ice? What do you think? Maybe it is faster.'

Mawson considers Mertz's suggestion. 'The trouble is, those coastal slopes will be even more heavily crevassed than the difficult area we just crossed. And there's no telling how broken up the sea ice will be.'

'Madi's men are on the coast . . . we might have a rendezvous.'

'Well, we simply don't know. I think it's best to travel back across the plateau. More inland. Further south. We should have smoother terrain. Less disturbed.'

Mertz nods, deferring always to the man with the greater experience of sledging in Antarctica. He is the leader. But somehow it is different. Now that they are only two men.

They're well matched, in that Mawson holds Xavier Mertz in the highest regard. Mawson is convinced he has the right man with him. Over the past month, Mertz has proven himself to be physically and mentally strong; with his mastery of skis and his affable nature, he is well equipped to endure the daily trials of polar travel. Neither is he afraid of hard work. There is plenty of that ahead of them.

CHAPTER TWENTY-EIGHT

They make a dismal procession. The dogs are already played out. Johnson barely stretches 7 kilometres before sitting in the snow, his sunken sides pumping with his sorry laboured breaths. Mawson urges him on, but the animal simply refuses to get to his feet. Mawson heaves him up, giving him the added motivation of a gentle shove forward, but Johnson sinks to the snow. He regards his master with a look of pathetic resignation.

'Let's get him on the sledge.'

Before the night is out, Mary and Pavlova have also lost the strength to be much use and are unhitched. How quickly the other dogs become disenchanted once one of their number has given up the fight. With fewer dogs pulling, the burden on the remaining dogs is far greater. Under overcast skies, the way forward seems dismal. They are frustrated by sastrugi: hard, wind-polished, like ranks of obedient soldiers intent on waylaying progress. The route zigzags. Both Mertz and Mawson trip over. Even the compass seems to have lost its way. It appears to have died from confusion.

A light snow falls. Visibility is reduced to a few metres; any definition is lost. It makes the going even tougher over the sastrugi. Occasionally a howl escapes from one of the dogs, a mournful protest that makes absolutely no difference to their plight. With Mertz out front attached with a tow rope and Mawson harnessed to the sledge with the two remaining dogs, hauling the sledge with the dog Mary barely alive onboard it must be an odd sight indeed. Yet they average over 20 kilometres a day.

Sometimes a boom reverberates below the snow like thunder. Conversation is clipped, everything that must be said is saved for later. And even then, they do not care for conversation. Mawson's eyes give him trouble. He thinks it might be conjunctivitis, but the eye drops in the medical kit do not produce any improvement. Bandaging one eye relieves some of the pain, but his sleep is disturbed. Perhaps it is stress. West is the direction of travel, but with no steering point and with the compass so useless, they could easily be deviating from their heading. Mawson worries that they have strayed too far south. There are no familiar peaks or landmarks, only hinterland, and a great deal of it. They throw away the alpine rope and the precious rifle.

'Xavier!' Mawson calls. They discuss a minor course correction.

Mertz is struggling already. He is snow blind and finds it hard to maintain a straight line with the constant weaving that is necessary through heavy stretches of sastrugi. It is agony to keep his eyes open. At times it feels like there is too much daylight; even at night, the glare is intense. The barest relief comes when Mawson instils the zinc and cocaine eyedrops.

They move through days of overcast skies over endless undulations. The sledge-meter freezes. In their efforts to fix it, their hands, made so clumsy by exposure to cold, snap the spindle. From now on they will have to measure the distance they travel by time. They agree that over a good surface they can haul at 10 kilometres an hour, over a bad surface, they advance at a third that speed. It is decided that further lightening of the sledge is required. Jettisoning the camera and other precious but ultimately useless equipment feels like a dreadful sin. To Mawson at least, still a man of science.

'That hypsometer was from the Shackleton expedition,' says Mawson with a sniff.

'Oh,' says Mertz indifferently.

Another dog is spent. Haldane. A big grey wolf of an animal. He has performed his duties valiantly. The last few days he has been too weak to even chew. How mournful he looked, holding the meat between his paws and merely licking it. And for what? A life of toil and mean living, then a not-so-quick death. Having left the rifle behind at their last camp, Mertz must dispatch the dog with a knife. He grits his teeth and completes the brutal deed. With a firm yet tender hand, he holds Haldane and waits for the dog's weak struggle to end.

Tears burn his bloodshot eyes as he is reminded of Ninnis's challenge to Bickerton. He had said: *'I bet my dogs will cover a greater distance than your plane.'*

And Bick had replied: *'Loser takes the winner to dinner at Trocodero then on to the theatre.'*

Will he ever see Bick again? Or go to the theatre in London, or see anything beyond this white, deathly place? Mawson has assured Mertz that he will see Australia and New Zealand. Mertz would settle simply for the hut at Cape Denison.

They cut up the carcass, consume a small portion of the stringy meat for supper and cook up the rest for later, as they've discovered it is lighter to carry cooked than in its raw state. The dog meat is foul in taste and texture. While it fills the gap in their bellies, it is so lean that it mustn't offer much at all in the way of calories. They toss Pavlova onto the sledge. Onwards. One dog remains. Ginger.

When hauling uphill, they make themselves ill with the effort. Their furious red faces strain out from their iced-up helmets, and the harnesses cut into their bodies. It's not just the dog in decline. Their own energy levels are depleted by the toil and a vastly inadequate diet. One night's supper is thin soup, made from simmering bones in a pot with snow. They cannot help but think of their own bones, which are starting to become more apparent with each passing day.

'Dog broth,' says Mawson with a grimace and one raised eyebrow.

'Did you wash socks in it?' Mertz asks glumly. 'It look like you did.'

Of course, neither man has changed his socks or underwear or clothing in the ten days since Ninnis died. Neither man mentions Ninnis. There seems to be a silent pact not to speak his name until they are out of danger. However, thoughts of him linger around the edges of every waking moment. And the thought of suffering a similar fate haunts every step. As far as crevasses are concerned, out of sight is definitely not out of mind.

CHAPTER TWENTY-NINE

What has become of their world? It is upside-down, and totally wrong. Mawson is up at 11 p.m. making breakfast. And it is Christmas Day. No turkey and trimmings; they'll be celebrating with dog stew. And very small helpings at that. Mertz and Mawson wish each other Merry Christmas *in the future.* This one is distinctly unmerry.

Mawson has started the day with a feeling of mild annoyance. He had been having a particularly pleasant food dream. He was in the presence of a huge cake, but woke before he could take a bite. Salivating at the mere thought of tasting it, he feels thoroughly betrayed by his imagination. His stomach aches in response. How he longs for something other than boiled dog. The stew will be hot, at least. And he does experience a sense of satisfaction in describing the cake in all its magnificence to Mertz, who listens intently and conjures his own version of the cake in his mind's eye.

'But what weird surroundings I found myself in,' says Mawson. 'I couldn't tell where I was.'

'No weirder than these surroundings I am sure,' jokes Mertz without laughing.

After they've scoffed the stew, Mawson tells Mertz that he has a surprise.

'I found these in the bottom of my bag,' says Mawson, opening his gloved hand to reveal a sledging biscuit, cracked in two.

It's the best Christmas present ever. He hands one half to Mertz and they nibble at them in a slow and deliberate fashion, enjoying the look of the friable texture as much as their mouths relish the feel of the miserable biscuit's small dry crumbs between their teeth. How strange to be absorbed in savouring something so meagre. Ten days ago they would have chomped it down in seconds without even noticing.

They talk for a bit about the strange sea bird that appeared out of nowhere. How it circled them for a while and then disappeared into the haze to the north. Both men recall the petrel that flew headlong into Ninnis's sledge but neither have the heart to summon the memory by discussing it or what it signified. It had happened only a day or two before the tragedy.

If only they could proceed in a straight line. If only they could fly! The hut is a mere 250 kilometres away. But avoiding crevasses and navigating through fields of sastrugi will see them covering a vastly greater distance. Eight to ten hours of intense hauling with only two men represents a great physical trial. However, when poor weather or lack of visibility makes travelling more difficult or dangerous, they must remain confined to the tent for a period of enforced rest. At such times the frustrations feel enormous. Their rations will only last for so long. *They* will only last for so long.

Ginger is the last dog to be slain. It takes Mertz an hour and a half to boil up all the meat from the carcass. The steam coats the roof of their saggy tent and freezes upon contact with the icy fabric. It lingers there like a mild ghostly reproach. The other things in the tent, the clothing, the sleeping bags, the men themselves, get progressively damper in the steamed-up space. Mertz in particular is suffering, having lost his overtrousers and his Burberry wind helmet with Ninnis's sledge.

Mertz is sick of living in such close quarters, of staring at the underside of their clumsily rigged tent. Especially on the long days when there is no point breaking camp due to foul conditions and they remain in their sleeping bags and drift in and out of sleep. He misses the booming sound of tent walls pulled taut. He hates waking to see the droopy fabric just centimetres from his nose. It reminds him of their plight. Everything leads him back to thoughts of Ninnis. And at times of inactivity, he is haunted by memories of his friend's sudden and unfair death. The injustice of it. The utter meaninglessness of it. The absurdity of it.

He had always enjoyed the sledging journeys, rejoiced in being released from the stale smells and confinement of the hut. Even when it required scuffling with the monstrous wind, he welcomed the openness, the limitlessness of the landscape. Now that he has to cross it, he curses it.

'We could rig sail from tent?' Mertz suggests during one such lull, when snowfall has waylaid their advance. 'Use the sledge runner for mast?'

'I think you're right. When the wind is a little stronger,' Mawson agrees. 'And blowing in the right direction.'

Neither man feels any real enthusiasm for the task, knowing that their first task will be to dig out the buried sledge. Rigging a mast and fiddling with the tent cover in order to devise a sail and attach it securely will inevitably involve an extended period without gloves on. And the thought of a belligerent wind without the protection of his Burberrys makes Mertz shudder with dread. They pray for a warmer day. Yet a warmer day will make the snow heavy going. They really can't win. This place is unwinnable.

CHAPTER THIRTY

Dawn is breaking. The sky is settled and pale and streaked with high cloud. It is very cold. The sun, having bounced along the horizon, is once again on the rise. In the soft light the ground is moving at high speed with panels of snow that skate across the surface, obscuring their boots and unsettling the gaze. It will take the two of them several hours to pitch the tent in this wind.

It's a tangle, a struggle, a five-man job at the best of times, requiring clear heads and good coordination. Instead, they must complete this seemingly simple yet fantastically complicated task in a state of high irritation, with frostbitten hands, tired eyes, empty stomachs and a wildly disrupting wind. First they need to unhitch it from the makeshift mast. What a useful piece of fabric – at once a means of conveyance and a shelter in which they can lay their weary bodies down.

They marched through the night and continued uphill, over a series of terraces spaced several kilometres apart. Now at the top of the climb, they're rewarded with a view of the coast. Clocking up 12 kilometres in total, they're satisfied with the sail, although when the wind is blowing a steady 50 to 65 kilometres the sledge is blown uncomfortably sideways, which makes hauling easier yet more complicated due to constant course adjustments.

Ginger's skull is a puzzle; they have no tool or instrument capable of dividing it equally. The brains are the reward they seek. Already the thyroid has been consumed – not a pleasant feast under any circumstances, but both Mertz and Mawson agree the wonderful bit is the sensation of having hot food sliding down their throats. They pass the skull from man to man and dig out juicy goodness by the spoonful, the organ's flavour and consistency being far preferable to the stringy meat they have been living on. They wash it down with a cup of tea, if one could call the tawny liquid that. It lacks any real flavour – this particular tea bag is on its last legs, after being reused and squeezed out four times already.

There have been days of indigestion. Days when they

both feel off. Yet something more serious is afoot. Mertz is not well. His stomach gives him trouble constantly. Laying off the dog meat, they allow themselves a small amount of pemmican and some cocoa from the meagre rations they carried on the first sledge. They can barely believe the luxury of butter, the way a square of chocolate melts on the tongue. It is a treat to celebrate the New Year. Yet all too quickly it is digested. But it does little to arrest Mertz's overall physical decline.

His lack of overtrousers has caused no end of strife. The fine wool fabric of his long johns, which is often iced up during the march, thaws and becomes wet in his sleeping bag. Never having a chance to dry out, his trousers remain damp; every time they set out, they freeze against his skin. The chafing that results when he walks has been excruciating. The skin on his thighs has rubbed off in places. He cannot seem to warm up, even on the march. The wind slices through him, his fingers ache with frostbite. By the end of the first week of January, his suffering is intense. Little by little his strength is ebbing away. As is his ability to do his share of the work – the breaking of camp, the cooking, the packing of the sledge, the pitching of the tent. Mawson suspects that even hauling the sledge will soon be too much for him to cope with.

'Overcast, a little snow falling,' says Mawson, coming back into the tent after surveying the weather and surface conditions. 'The wind has died down too. We should try and get going. I think it might clear.'

His prognosis yields no response. Mertz has slipped back down into his reindeer sleeping bag.

'Xavier? Did you hear me? We should get going.'

'I can't.'

'Come on, I'll help you.'

'I can't.'

Mawson inhales deeply. 'Xavier, we must travel when we can. We must make up distance. Even a few miles. For exercise. Come on now.'

'Suicide.'

'What?'

'It is suicide.'

'That's not the attitude. You must move. We cannot stay here. We must use every day – as much as it is feasible – to cover some miles. Our supplies will only last so long.'

There is no movement from the sleeping bag.

'Xavier, we must get on!' Irritation crackles in his voice.

'It is best for me. I need to stay in my bag today. I need to dry it. I need to get better.'

And so they stay.

'Tomorrow,' says Mertz. 'Please.'

Mawson closes his eyes and nods.

CHAPTER THIRTY-ONE

Mertz has a fever. Still they break camp. It takes them three hours to get underway. The surface is slippery, it chips away at their resolve. Both men feel dizziness from having spent the last thirty-two hours lying down in their bags. Weak from hunger they stumble, trip; occasionally they fall to the ground.

'Xavier, are you all right?'

'I must take a pause.'

'But you took a rest barely ten minutes ago!'

'Give me a moment.'

Mawson is silent with rage when, after barely three kilometres Mertz refuses to travel any further. He offers to pull him on the sledge.

'No.'

'Come on. I shall rig the sail.'

It takes some time for Mawson to arrange everything with Mertz looking on, a forlorn figure. He is completely childlike in his helplessness and lack of self-consciousness. When he climbs onto the sledge, it is with a sort of mute indifference. Such an attitude is entirely out of character. It does not bode well. Sure enough, before long Mertz expresses a strong desire to halt and pitch the tent. It is the wind, he says, it is cutting into him like a razor. Mawson feels like shouting that it is cutting through him too.

Single-handed has become the norm; as has frustration. With significant difficulty, Mawson pitches the tent and hurriedly prepares some food. To his horror, he realises that Mertz is not only reluctant to eat, he seems incapable. Food will not pass his lips. Mertz folds himself awkwardly into his bag and starts to tremble as his eyes close. Sleep claims him almost instantly.

Lying awake, Mawson begins the calculation. Always the same, always involving the same horrendously inadequate variables. Food, distance, time. They need to cover 15 kilometres a day if the food is to last. They cannot afford any more rest days. Even another couple of days of inaction and they're doomed. But how can Mertz progress in his current brain-addled and weakened state? He is likely to deteriorate even further if he refuses food. Not only does the future look grim for Mertz; it looks exceedingly grim for Mawson.

Flakes of snow thud on the canvas over their heads. Every now and then there is a gust of wind that sounds like a sudden coughing fit. Mertz gives a groan and Mawson tries not to linger on dark thoughts. Both their chances are fading now. Each time it occurs to him, Mawson chases the same monstrous thought as far back in his mind as he can. It is getting harder to keep it locked away – this awful, awful thought . . . of going it alone. He cannot leave. He mustn't leave him.

CHAPTER THIRTY-TWO

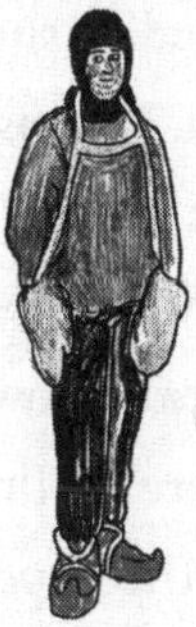

It is 8 January. Mawson rouses his companion from sleep. He coaxes Mertz from the sleeping bag and attempts to dress him as quickly as possible. Mawson realises that Mertz has soiled himself. It is a sorrowful situation, yet Mertz scarcely notices as Mawson begins the awful task of cleaning him up. It takes a long time. The smell is foul. Death is stirring.

As gently as he possibly can, and taking care with his poor raw legs, Mawson rearranges Mertz back into the sleeping bag. He has become so cold himself that he too must slide into his bag in order to bring some warmth and movement back to his frozen, rigid limbs. Two hours he lies shivering, his thoughts circling around the notion of survival.

The *Aurora* is due back at Cape Denison in a mere six days. It may already be there. Perhaps they should have proceeded north to the coastline. He had left instructions with Captain Davis that should he not return, the ship might be navigated eastward to look for them. And what of the other parties? What has become of them?

Mawson feels wretched at the thought of the lives put in peril for the expedition. Ninnis gone; Mertz teetering on the brink of death. What other casualties have there been among the other parties? Wild was right. The men were inexperienced, lacking any real knowledge of polar extremes. He catches himself before guilt drags him downwards. Each and every man on the expedition knew the dangers of polar exploration. Each man chose to be here. Ultimately there was nothing about Ninnis's accident that could have been avoided. They had all suffered innumerable falls into crevasses, but never anything so severe as to cause injury. He doubts that complacency played a part.

He shudders at the thought of Ninnis's family hearing the tragic news of their son's fate. And his heart aches when he considers Mertz's kin - unaware that their son and brother is dying on a most desolate plateau on a far-off and deadly continent. And poor, poor Paquita. Of course his fiancée knows the risks. They spoke of the potential for disaster before he left Australia, but will it make news of his demise easier to bear? Mawson thinks of the letter he penned on leaving the hut, in which he told Paquita that if something should happen to him, he wanted her to know how much he loved her.

Mawson shakes himself from sombre thoughts. Ruminating on things he has no power over is pointless. He says a silent prayer – the help of the Almighty is what he needs. He lights the Primus and mixes up cocoa for Mertz, who is incapable of lifting his own head to drink. Propping up his friend, he encourages him to take the liquid. The smell that rises from his sleeping bag is atrocious. His dysentery has worsened. Mawson tries not to gag while cleaning him up. What else can he do? There is no place for annoyance or self-pity. These are surely Mertz's last hours and he will do all he can to make him comfortable. Already his companion is slipping into some form of delirium. He cries out and thrashes about, breaking one of their tent poles in the process.

When the fitting commences, Mawson is at a loss. Should he restrain him? He holds him down and waits for calm to take hold. Mertz shouts and rages, he babbles and slurs. His words are loose, some English, some German. They don't string together but rise and fall in fathomless rounds of nonsense. Trailing off to sleep, he wakes with a shudder and starts again with an incoherent warbling.

'O weh, o weh, o weh,' he cries for what seems like hours on end.

A fever rages on his brow. Another drink is perhaps what is called for. Yet when offered one, Mertz will not take it. He refuses everything. He falls silent. Peaceful in sleep, finally, he dies.

With considerable reluctance, Mawson draws his hand from his own sleeping bag and places it on his companion. Hoping for a vague stirring, a feeble murmur, he finds the stillness of Mertz's body simply dreadful. He withdraws his hand. He blinks. His mind hovers over this man, whose generous spirit contributed so much to the expedition, to this sledging journey. Yet his cheerfulness and robustness was not enough. Mawson half expects to feel warm tears start down his face, but none arrive. He should cry for Mertz, he really should. Equally there could be tears of self-pity. Reaching the hut seems an impossible goal. He can barely summon the belief that he'll be able to break down the tent and load the sledge single-handedly.

Mertz is dead, he says to himself, horrified at the finality of the words. *Mertz is dead.*

CHAPTER THIRTY-THREE

Lying in a damp sleeping bag for a week and barely eating has laid waste to Mawson's overall physical condition. An unforeseen effect of the poor diet is that the body refuses to heal. Even with enforced rest, his skin has erupted in sores. His lips are split and his nose is a mess of wind-ruined skin, peeling and inflamed. His legs and scrotum, rubbed raw, are every bit as bad as Mertz's. It is painful to walk, but also just to hunker down in the bag.

This past week has seriously eroded his mental reserves. Anxiety stalks him now. Forced to remain here, waiting for Mertz to rally his strength, has very probably scotched his own chances of survival. Weakness is creeping into his limbs. He feels every fibre of his muscles strain against the task of hauling Mertz from the tent. Of course, it is no longer Mertz. It is merely his body, cocooned in his spoiled bag.

Mawson is momentarily bent double by the effort; he is left breathless with emotion. He collects himself, inhales deeply and looks out over the landscape and wonders what are his chances. There are 150 kilometres left to travel. He needs to do it all on his own, without the morale-boosting presence of another; of Xavier.

The sledge is too big. It is a great heavy beast of a thing, even without essentials loaded onboard, it weighs a tonne. It's not just that he's sick of pulling it, he is incapable. It occurs to him that maybe it could be cut down quite successfully. The difference would be significant – even a question of life and death.

The next day there is the ghastliness of having to bury his companion; of having to once again read the burial service. Mawson goes through the motions. But there is also a stirring of his spirit. The act of remodelling the sledge focuses his resolve. He makes modifications to the mast and sews a new sail from the clothing bag and Mertz's Burberry.

How easy it would be to surrender to the situation and simply retreat to his sleeping bag. Death would come eventually, and all of this suffering would be over. Regret clusters around his inability to share the survey. They've mapped out over 500 kilometres of coastline, with significant notes on glaciers and ice formations. And all the information that he has committed to memory will be lost if he fails to reach help.

At least his diaries and those of Mertz will tell the story of what happened. That in itself is worth continuing for. He feels he has a purpose. Survival might not be possible, but he has his purpose now – to get somewhere his body might be discovered. In accordance with this realisation, Mawson prepares himself a fine evening meal, allowing a far more generous helping of the rations than he would have had otherwise. He needs his strength.

The following day his determination is tested. The blasted wind is back and putting up such a fight that setting off is out of the question. He'll never manage to pitch the tent again in such a horrendous howler. Without securing shelter he'll die. Outside has become a skidding, clobbering assault of horizontal ice particles. The ground is a white shroud, spreading ever outwards – ready to receive him.

CHAPTER THIRTY-FOUR

It feels like his body is decomposing. The sores on his fingers weep and fester. Skin is beginning to peel off his legs and torso in strips. His nasal passages feels like they have been seared raw and his mouth produces so little saliva that he finds it hard to swallow. He supposes this must be because he is not eating well. His system is not getting the nourishment it desperately needs.

Although the weather conditions have been favourable enough to set a course and begin the march, his feet feel increasingly uncomfortable. It's an odd sensation. They feel somehow lumpy. Even with this impediment, he manages to achieve a distance of 10 kilometres before giving in to the pain. Taking off his finneskoes, he steels himself to what he will find.

The state of them! It is a hideous discovery. The soles of both feet have formed enormous blisters full of fluid. Clearly, while walking on the blisters, he was spared the worst of the pain; now having burst, these blisters offer little protection for the raw skin beneath. Walking on two fresh wounds is what he has been doing for the best part of the day. How long will it take them to heal? He feels utterly crestfallen. It is another nail in the coffin. Had it not been for the sun, he might have succumbed to despair. It is a beautiful day, warm and windless. Mawson exposes his sore feet to the sunshine and rejoices in the comfort it offers, its rays feel like medicine that is pure benefit. Stripping off his layers, he bares his body and can scarcely believe the good it does him to lie for a few short moments in the bright sun. It's not merely the physical sensation; it's a balm for the soul. It restores him just a little bit, feeds him just enough.

The fine weather does not last, however. Barely twelve hours after his glorious sun bath, Mawson must confront another day of strong winds and driving snow. Laid up once again in the tent, Mawson accepts that it is probably the best outcome. Keeping off his poor mangled feet for even a day will be a relief. Instead, he makes a study of how much food he has left in the ration bags to augment his reserves of dog meat. Since Mertz's deterioration, he made a practice of picking bits and pieces out of the week's ration bags – the choicest bits, using up the things that he felt would be most palatable. To keep up Mertz's flagging spirits. What's left is an odd assortment of items. When it is laid out before him, it's a woeful spread. If he had twice as much, he would prepare a fine meal to lift his own flagging spirits. Food has become his focus.

Breakfast is one sledging biscuit, half a serving of pemmican with a little shredded dog meat and a cup of coloured water that scarcely passes for tea. The day is spent dozing, and periodically shaking the snow off the tent, which, in its frightfully saggy state, ends up bowing inwards with the weight. Amazingly, the afternoon clears. Taking his chance, Mawson sets out across a wide glacier.

The icy slopes descend over crevasses, but despite successfully navigating the hazards, the hard and knobbly surface is murderously painful on his feet. He cannot do more than 8 kilometres. He halts. The feet are in an even worse state of repair. They throb, they weep, they sting. But he must persevere. Come morning he will press on.

Undoubtedly distracted by the deplorable state of his feet, he forgets to wind the chronometer in the evening. It is a silly mistake. It seems astonishing that an action that was performed day in and day out, every evening for so long; indeed, like *clockwork*, could just drop from his memory. It is not until the next morning that he realises this dreadful mistake. His watch has stopped at twenty minutes to 11 p.m.

Without an accurate idea of the time, he cannot make calculations based on his tables listing the position of the sun. Without this, he cannot establish his overall position. How else will he navigate across this vast white wasteland?

Everything is now slightly off. His watch certainly is. Since the tripod legs of the theodolite have been splinted together and used for a tent pole, his sightings have been taken by balancing the instrument on the cooker box on the sledge. So it is not exactly accurate. Even the distance he has travelled is an approximation. With mildly inaccurate sightings taken with a theodolite without a tripod, and in poor low light, he is not entirely confident he has it right. Going by dead reckoning, he is heading in the right direction. And in a straight line. Then again, with his feet so bad, his chances of getting back to the hut are diminishing considerably. And one thing is certain, he cannot fly home. Would that he had Bickerton's *Grasshopper.*

CHAPTER THIRTY-FIVE

The mood at Cape Denison on 17 January is lively and jubilant. To be back at the hut and sitting among friends around the table gives the purest pleasure. To still be alive is another source of joy; they have all had close shaves with death. Yet nothing can compare with arriving back at the hut to find the *Aurora* anchored offshore and Captain Davis with his mail sack bulging, and neatly bound bundles of correspondence from friends and family.

Once news from home has been devoured and the longed-for letters have been read and re-read, and folded up and then opened again for yet another reading, the men turn their attention to the newspapers, which are full of the most eagerly awaited reports concerning the race to the South Pole. It would appear that Roald Amundsen and his men were victorious, having reached the southernmost point on the planet on 14 December 1911, more than a year ago. There is as yet no mention of Captain Scott, who had not yet returned from his journey when the relief ship picked up most members of his expedition earlier in the year.

Reading about Amundsen's use of sledge dogs, it gives the men a thrill to think that twenty of the Norwegian explorer's dogs are presently outside their own hut. Having called into Hobart on his voyage back to Norway, Amundsen knew he no longer needed the dogs. Rather generously, he offered them to Captain Davis. Are the dogs happy to be once again on the ice? Madigan spares a thought for the poor blighters – he cannot wait to leave this place and its wretched wind.

It's an evening of bluff and bluster, talking together, sometimes wild shouting over each other, everyone sharing experiences both humorous and alarming, and the discoveries and disappointments that will forever unite them. Now that the end is in sight, that they are sure of being able to leave very soon, they feel they have been part of something larger and worthwhile. The feel like a team.

They raise a toast to sledging rations being a thing of the past. They tuck into fried seal steaks and gorge themselves on chewy slabs of bread, slathered with butter and jam. The thought of a real bed is just too good to contemplate.

As always, Hurley is full of the best stories; the highs and lows of their southern adventuring sound like a Boys' Own tale. He, Bage and Webb ventured an astonishing 500 kilometres from the hut, yet were forced to return before reaching the magnetic south pole. At times they were unspeakably cold, snow blind, frostbitten, almost ready to die from the despair of another day staggering through hideous windblown drift. The tent offered no respite – the inescapable freeze tortured them even in their sleeping bags. On the return they almost starved.

'Bage measured everything out, made sure everything was scrupulously fair.'

'We always felt we were the one missing out, of course. And we took the greatest pleasure in savouring every last crumb.'

'We realised our food wouldn't last if we kept going,' says Hurley. 'But we pushed as far as we could. And we did cut it a bit fine . . .'

'Almost finished us, didn't it?' says Webb.

'Yeah, but it didn't. And I reckon it's a new world record.' Hurley's enthusiasm is irrepressible.

Says Webb, 'I tell you, reaching Aladdin's Cave was the most memorable day of my life. All that food! Not that Bage's eyes could take any of it in.'

'They had virtually closed up by then,' says Bage. 'I thought I had ruined my sight forever.' His eyes are still a mess. His snow blindness was undoubtedly worse than anyone else's.

Madigan, Correll and Dr Mclean have had their own outlandish experiences – times when they were exhausted, weak and chilled, times when the smallest detail was a source of joy – the plum pudding made by McLean – a thick mixture of ground-up biscuits, raisins, sugar and milk powder, stirred together with whisky from the medical kit.

'No one would thank you for it now,' laughs Correll.

Says Madigan, 'It was the hardest Christmas I have ever had. And the most wonderful.'

His coastal team have had their fair share of drama, choosing as they did to travel over the sea ice instead of land during their eastern coastal survey. The surface was far easier to travel over than the rutted, often icy coastline. Yet a few times they didn't anticipate that tide cracks might make the crossing difficult and that there was a risk of being washed out to sea on an ice floe. Another time, when camping, Madigan's body broke through the surface to his armpits; they then realised to their horror that they had slept the night on ice that was only a few inches thick. Trudging along in clothes soaked in seawater, with only the wind and body heat to dry them, will go down as one of Madigan's least favourite Antarctic experiences.

There are other tales to tell. Laseron recounts how Close saved his life and that of Stillwell. Sheltering in Aladdin's Cave during a blizzard, both Laseron and Stillwell had lost consciousness due to carbon-monoxide poisoning. 'The entrance was blocked up with snow,' he explains. 'Close was also pretty close to fading, but he immediately understood what was happening and thrashed a hole in the accumulated ice and snow with his ice axe.'

A few of the men slap Close on the back and say, 'Good job, Terribus.'

Close just shrugs and is glad he seems to have restored his reputation after his near-total collapse during the sledging journey with Madigan and Whetter.

Bickerton's team are the second-last group to return. They stumble in through the door just after midnight. Entering the hush of the hut, amid sounds of snoring, Bick takes great relish in calling in a booming voice, 'Rise and Shine!' – and seeing all the sleeping bodies leap from their beds in fright.

They were not so successful with the *Grasshopper*. In fact, the air tractor turned out to be a complete flop. After just one day, the engine seized so abruptly that the propeller shaft cracked and the propellor itself became unusable.

'We left it where it died. Pulled our own gear after that,' says Bick. 'Four sledges.'

'All hundred-and-eighty kilos of it,' says Hodgeman, narrowing his eyes and grimacing.

'In eighty-kilometre winds,' adds Whetter.

'Wasn't too bad,' says Bick. 'Alfie kept us entertained with his sketching. And he even organised a tree for Christmas. He drew it on the frost that lined the inside of our tent.'

Given the hundreds of hours he had spent adapting the *Grasshopper* to sweep across snow instead of fly through the air, one might imagine Bickerton's devastation when confronted by such a spectacular failure. But he doesn't seem too put out. Not a day later, and trudging along on foot, Bickerton made an incredible discovery – a one-kilogram rock sitting in the middle of nowhere on the top of freshly fallen snow. It was a perplexing sight at first.

'I found a meteorite,' he says animatedly to Madigan as he retrieves it from his kit bag. 'Can you imagine this careening through outer space.'

'Can you imagine it hitting you on the head?'

'You wouldn't be capable!'

Bick places his trophy on the shelves at the head of his bunk.

'It might turn out to be the first one ever discovered in Antarctica,' says Madigan.

Bickerton gives a satisfied sigh as he stretches out full length on his blanket. Can there be anything more wonderful than the thought of a peaceful sleep, of not having to wrestle with a tent in the morning or eat pemmican? A comfortable night out of the wind, with blankets instead of frozen or damp reindeer fur?

Madigan is overjoyed to welcome back another member of Hyde Park Corner.

'Still not back, those two?' Bickerton murmurs as he drifts off.

'Not yet,' Madigan says. But he has no doubt it will be an even more joyful reunion when Mertz and Ninnis join them for their final few nights in the windiest place on earth.

CHAPTER THIRTY-SIX

I'll be overdue now, Mawson thinks. By several days. Everybody was to be back by 15 January. Will they send a search party out? He can only pray. Every mile he struggles on is hard won. But he might be able to get to Aladdin's Cave. There, where there is an abundance of food, he can get out of the wind and at least recuperate until his feet have improved. But what of the *Aurora*? Has the ship returned? How long can Captain Davis be expected to wait? What if he is assumed dead, and the ship departs, stranding him on the ice for good? He pushes it away; the thought is too horrible.

Over the last few days, his feet have got much worse. Every step sends a knife-blade of pain shooting up his legs.

The only thing that softens the edge is when he sits down, slips his boots off and exposes his sorry feet to sunlight. That is, when there is sun. The overcast days offer good travelling, at least. When the sun beats down, the surface deteriorates, melting the snow to slush. It is like pulling through deep sand.

The rotten condition of the snow makes hauling the sledge impossible. Exasperated, he plonks himself down to rest. He must decide whether to keep on for a few more agonising hours, or call it a day and go through the effort of pitching the tent when he has made precious little headway. Removing his boots and peeling off his socks, he makes a gruesome discovery. His socks are damp with blister fluid, his toes are black and festering, and the entire sole of his left foot is coming away. Gulping, he examines his right foot only to find the same thing. He closes his eyes.

Carefully he eases the skin back and stares dumbfounded at the leathery flap he holds in his hands. The dried thickened heels are the colour of candle wax. He barely comprehends – they do not seem like anything that was once part of his body. The new skin underneath is pink, weepy and raw. They start to sting simply on contact with the air. How can he bear weight on them? The tender skin is certainly not ready to meet with a sock or even encounter the ground. How on earth will he manage to walk now?

There is only one thing for it. From the medical kit he retrieves a roll of bandage. He positions each flap snugly over the sole of each of his feet and, gritting his teeth, he winds the bandage around and around. Extra socks will provide a bit of cushioning. He's going to need to wear every pair he has. Consumed by his immediate task, he remains economical with his thoughts of what lies ahead. There is no room for self-pity; neither does he see anything heroic about his journey. He must apply all his energy to the business of survival.

CHAPTER THIRTY-SEVEN

The relief party consists of Hurley, Alfie and Dr McLean. On gaining the blessed safety of the hut, they had all vowed never, ever to return to the blizzard-cursed plateau. Were it not for the rising concern all of them feel for Mawson's party, they would happily have boarded the *Aurora* and sailed for civilisation without a backwards glance. But duty calls and they have all willingly risen to the challenge.

In Mawson's absence, Gloomy has appointed Madigan leader of the Cape Denison hut and issued instructions. Hurley, McLean and Alfie are to walk inland to the south-east and erect a series of snow mounds, where they will leave rations and dog food and sealed tins with notes outlining clear instructions on the distance and compass directions back the hut.

Mawson, Mertz and Ninnis are ten days overdue, and they will be running seriously short of provisions by now. What has waylaid them? They had all the dogs, so it should have been reasonably fast on the return.

'I cannot imagine for a minute that Mawson would have taken any risks,' says Hurley.

'Or lost his way,' McLean points out.

Maybe one of them suffered an accident and they have had to haul the injured man on the sledge, like when Hurley and Webb had to haul Bage when snow blindness temporarily robbed him of his sight. That kind of toil could easily add a week to their schedule, especially if they've had to resort to eating the dogs. Rotten weather to the east may have meant that they were obliged to remain in their tent for a week, not wishing to risk it. There is little chance that they starved. They could always consume the dogs. What else could have transpired? Lost their way? Poisoned themselves with spoiled rations? It is impossible to guess.

'Make sure you're back here in five days,' Madigan says. 'The *Aurora* will be conducting a search along the eastern coastline, but when Davis gets back, he'll be keen to leave as soon as possible. He still has to get further west for Frank Wild's men.'

Captain Davis has explained the situation with Wild's western party, how he had to drop them on a highly unstable glacier tongue 1900 kilometres to the west of Cape Denison.

They want to maintain an air of calm. They have no desire to unsettle anyone, but Madigan and Davis have discussed the unthinkable. Even though the idea is abhorrent to him personally, Madigan knows that if Mawson's party do not reappear, a number of them will have to stay on for another year at Cape Denison. The thought that it will need to be him, as interim leader, fills him with grief and frustration. Another whole year! Not to see his beloved Wyn; to sacrifice another year and to put his precious studies on hold still longer. He prays that Hurley and his team are successful in locating the overdue ones – alive or dead. How he longs to leave this cursed place.

CHAPTER THIRTY-EIGHT

It's the second day of beautiful, obliterating, clarifying sun. Hurley, McLean and Alfie are grateful that the air is calm and that on the plateau everything appears still and serene. Building snow cairns to aid navigation seems almost a pleasant task like making a sandcastle. On such golden days one might be forgiven for assuming this is the true nature of the Antarctic continent, under the thick layer of bad weather. But they've all experienced enough of the polar environment to know that its true character tends towards cruel and tempestuous. This lull will not last long. In fact, only a matter of days from the hut, conditions deteriorate. A blizzard blows in. Ah, the Antarctica they all know. And hate.

Despite the foulness of the weather, the temperature remains warm. The blizzard pelts them with heavy snowflakes that splatter on contact. Before long they are soaked to the skin, their clothing absurdly wet, hanging off them in heavy layers until they decide they've had enough of this tropical parody of polar travel and will camp for the night.

They pitch the tent and take turns rolling back and forth across the groundsheet to tamp down the snow to a uniform flatness. It is easier said than done, and they end up with a large dip in the middle, which will mean that once they spread their equally soaked reindeer sleeping bags, all three of them will end up squished together. This might not be such a bad thing, knowing as they do that the moment they stop moving about, their wet clothing will start to refreeze and they will get colder and colder.

The next morning there is no rejoicing at the state of the surface. The snow is a soupy, slushy, crumpled mess. The sledge does not glide. It must be dragged. The thick layer of lumpy ice that forms on the sledge runners regularly needs scraping. Thankfully their load is not horrendously heavy. They all prefer being on the move to shivering in place.

'Things could be worse,' Hurley says with a grin.

They soon are. The wet snow is unstable underfoot. Their legs break through, they catch themselves continually; sometimes they slip to a harness length, shouting a warning to their sledge mates. Hauling each other out of increasingly sticky situations is exhausting.

By the fourth day they are at the limit of their search.

'I think this is it,' says Hurley. 'The end of the line. And not a sign of them.'

'Perhaps we missed each other, passed them in the blizzard,' says Alfie. 'We could have been within twenty metres and simply not seen them.'

Together they cut blocks of snow and pile them high in the last of the cairns. They cache the last of the supplies, keeping just enough to get them back to Cape Denison. It's a bittersweet moment, turning for home.

The weather is clearing. The wind is freshening from the south. The return journey will be quicker – travelling with the wind, they'll make rapid progress. They're motivated too – the prospect of getting dry is unbelievably appealing. Each day they have peered into the gloom, and now that the weather allows, they can see a full five or six kilometres into the distance until mist robs the horizon of any definition.

Hurley lifts the binoculars and gives one final slow sweep of the vast plateau. He lowers them. Looks at his two companions and simply licks his lips.

'So that's us then. We've done our best,' says Dr McLean as though the words actually needed to be pronounced before they could move off.

'May the Lord preserve them,' says Alfie.

CHAPTER THIRTY-NINE

The records are in order. Of this Mawson can be proud. Throughout his ordeal he has noted down wind speed and direction, temperature, cloud patterns and the condition of the surface, his position and the distance travelled – as close as he can estimate it. Should he not pull through, he hopes his corpse will be found and then at least his logs and his scientific notes will be saved and may be put to good use.

He thinks often of the expedition and its goals, how it will be perceived. Undoubtedly the deaths of Ninnis and Mertz will overshadow their many achievements, the new knowledge they have all contributed, and of course the coastal survey. And what of his own death? How will he be remembered? What will his legacy be?

He has written up his diary at the end of each day too, even though he scarcely has energy for it. Its pages will tell the whole desperate story of what happened to Ninnis, how Mertz was not able to survive. And at least his own final weeks will be set out in clear detail – how he tried to get through, how he really tried. To his very last breath.

He thinks of the photograph of his return to Adelaide after the *Nimrod* expedition, how he was pulled along the street in front of the university by students and his colleagues. How everyone suddenly sought his opinion and how he had been asked repeatedly to lecture on polar exploration. It had given him the greatest pleasure.

'Would you like to go to the Antarctic again Dr Mawson?' people had asked him.

'Never,' was his ready answer.

But how wrong he had been about that. He imagines them asking a different question: *Would you like to get home Dr Mawson?*

Yes! Yes! Yes! he thinks. Whatever it takes.

CHAPTER FORTY

The tumble catches him completely off-guard. He's brought up short by his harness and the air is knocked from his lungs. He swings then spins as the rope pulls taut. A waterfall of snow cascades onto his head. Winded, he tries to steady himself on the walls of the crevasse. What just happened? All of a sudden he's dropping, sliding ever deeper with bigger chunks of snow thumping down. With horror he realises his weight is pulling the sledge closer to the edge. It won't hold. It can't possibly hold.

Squeezing his eyes shut, Mawson braces himself for the inevitable. *So this is the end*, he thinks calmly. *Please let it be quick.*

But the sledge does not crash down on his head. A few unsettling moments pass. He slips no further. He suffers a light sprinkling of icy particles from above but nothing more. All is still. It's a horrible sensation to have so much snow down his neck. The rope squeaks in protest as it stretches; it is an eerie reminder of how precarious his situation is. It might snap. It could snap. He realises his eyes are still firmly shut. Finally, he dares to open them.

First he looks up towards the light. About four metres above him, a ragged-edged hole reveals a patch of blindingly bright overcast sky. The surface. It is a long way up but still, it is within reach. With his heartbeat still thumping in his ears, he dares himself to peer down. His eyes take a moment to adjust. Below his dangling legs the abyss soon swallows all the light filtering from above. The black bottomless depths stare back.

He stretches his hands up. By pushing himself off the sheer sides of the crevasse walls he manages to grasp one hand around a knot on the rope. Again he braces for the sudden cold rush of the collapsing snow lip, but it holds. Adrenaline kicks in. From nowhere, from somewhere, from his wizened, weakened frame: a surge pure of energy, enough to heave him upwards. He grasps another knot and the next. Within moments his head appears at the surface and he is breathing in the open air and hauling himself beyond the opening and smiling. With a whump of collapsing snow he's back hanging in his harness in the void, his head full of jarring rageful failure. His attempt has robbed everything from him; his mental and physical reserves are exhausted. Hanging his head, he catches his breath.

Suddenly he is conscious of how intensely cold he has become, with his clothing so full of snow. How sick of this he has become. To fall would be to disappear. A moment of rushing air then oblivion. Then, perhaps like Ninnis, his experience of all this would be at an end. A life extinguished in an instant. The pain of his ruined feet gone, the agonising raw skin on his legs and scrotum would be a thing of the past. The thought hangs suspended in his mind. The sweet brief temptation of it spinning slowly. But what if he should slip out of his harness, surrendering himself to the bottomless crevasse only to hit a ledge on the way down where, broken and bleeding, he must wait for his injuries to carry him into the next world?

Then the thought is replaced with another one.

Think of the food.

He simply cannot bear it. There remains precious food on the sledge. Who will eat it? Nobody will benefit from it. And the sun. How desperately he would love to feel its warm caress on his body. And Paquita. To never see her again – how can that be? And his scientific records. Why has providence spared him thus far? Why if not that he should continue? What he must do is find the last ounce of strength and push everything upwards. He has done it once before, he can definitely do it again.

The greatest effort unites his body and mind. Upwards, upwards. Now feet first, he wedges himself in the hole and levers his torso out. Grunting. Up and out and over onto his back. Wheezing, panting, cold oxygen filling his lungs to bursting. The air, the sky, the sledge.

CHAPTER FORTY-ONE

He wakes to find the day has slid away. His lifts his gaze. The snow around him shows evidence of his exertions. The sky is low-slung and pale and wholly uninterested in his miserable plight. How long has he been lying here, all strength gone? He is wet and cold and empty-headed. Remaining prone makes no sense; not after that struggle. *Collect yourself*, he thinks – then rolls over, gets onto his hands and knees, moves over to the sledge, unlashes the cooker, lays one hand on the food bag, heaves himself to his feet.

It takes an age to erect the tent. A little food passes his lips but it is as if he is still too rattled, too fretful to gorge himself as he might – as he had promised himself he would while dangling above death. He is lying in his reindeer sleeping bag, but sleep does not come. Instead his mind ranges over the day's exhausting events and the very real prospect of it all happening again. How grim. In his weakened state, he's unlikely to have the strength to do that a second time. A few lines of poetry arrive unbidden:

Unborn tomorrow and dead yesterday,
Why fret about them if today be sweet?

So he could just lie here, rest his feet, eat his fill, forget all the anxiety associated with eking out his severely depleted provisions. To spend his last few days without gnawing hunger pangs would be bliss – would that not be preferable to stumbling into another crevasse and coming to a violent end or dangling without any means of escape?

Another thought enters his head. What if he were to make a sort of rope ladder and attach it to his harness? Then he would be able to climb out. Assuming the sledge did not plummet with him, of course. Yes, it is an idea worth pursuing.

All night the wind and drift torment him. The cracking and booming of the glacier beneath him also play their part in depriving him of sleep. In the morning, he works on his rope ladder. By 8.30 the clouded sky and flat light make way for the sun, but all too briefly. Every step becomes a lottery in such poor light when the morning's route lies through a maze of huge crevasses. His safety device might soon be put into service.

The zigzagging between the fissures and cracks is tedious and adds mileage, but it is entirely necessary, until he discovers that thankfully these crevasses appear choked with fresh snow. He is delighted that he can log a distance of almost 9 kilometres a day, but every day brings near total exhaustion, with very little food to bulk up his energy reserves. If only he could get some sun. And some proper food. His body is breaking up. He no longer recognises it as his own. His fingertips are black and festering, his beard is coming away from his face, and his hair is coming out in handfuls. Strands of it float on the top of his pemmican, along with the spikes of reindeer fur from his sleeping bag, which seems to have given up. Running his hands over his scalp, he can feel great bald patches opening up. His feet never feel warm; the hot drinks never make it that far down his body. His feet, his poor feet.

Day in and day out, he steels himself to the task. He must keep a good pace. Seven kilometres is his goal. Although he can realistically only manage five. Through wet snow. Deep snow. The painful boils on his legs need lancing – it is rarely possible. The wind picks up. Were there still two of them, they might rig the sail.

He needs a lighter load. Unnecessary things must go, he decides. His crampons are tossed away. The wind falls off. The sun gleams through the cloud cover. Skin peels off his hands like a sheet of fine paper. He eats one whole biscuit and a decent, properly brewed tea. The experience is transcendent, yet soon a memory.

A blizzard hits with violence and keeps him in the tent. The deep snow, the soft snow, the heavy snow. It rains down, bombards him. His shelter droops under the weight until it is nothing more than a coffin-sized space. Mawson breathes in and out clouds of white, convincing himself that he will not suffocate. In the morning he leaves the tent to find the sledge has disappeared under a drift, while his legs disappear to the thighs. No progress will be possible in such circumstances. Another day wasted. He clears the sledge runners.

He thinks of Paquita. He sleeps. He frets about how long his oil will last. He dreams of the richest foods, offered to him in such quantities that he salivates in his sleep and cries out in pain when he wakes to find himself once again in frozen desolation. Confronting yet another day of the starkest privations transforms his dream to a nightmare.

The uphill pulling very nearly ends him. There are terraces that take all his mental resolve to clear. Yet the downward slopes leading him off the glacier make handling the runaway sledge almost impossible. He throws away his alpine rope. Reluctantly he abandons the theodolite. Breaking camp, he leaves it on the snow. A little dark instrument amid a white desert. From now on, he'll rely on the compass alone. Everything is saturated with melting snow, with perspiration, the fluid from weeping wounds. He fortifies himself with an extra stick of chocolate. It fills his mouth with sweetness, richness, goodness. Then, like a dream, it is gone.

CHAPTER FORTY-TWO

Madigan writes letters. The most important one is to his fiancée Wyn, but there are many others he wants to write to explain why he will not be returning as he had desperately hoped. He was so sure of leaving aboard *Aurora* that he burned quite a few of the letters that sustained him during the year on the ice. Convinced as he was that he would see his friends very soon, he had no need for the newsiness of them that had grown terribly stale from reading and re-reading.

Bick, McLean and Alfie have offered to stay on with Madigan. Bage too has elected to stay and is busy writing a letter requesting another year of leave from the armed forces, which he intends to send back with Captain Davis when the *Aurora* sets sail.

'What if they refuse to grant your leave?' asks Madigan.

'Then I expect they will reply by return post,' Bage chuckles. 'And once I receive it next January I shall resume my army duties.'

There is a new man, Jeffryes, who has taken over the wireless operations from Hannam. With the benefit of more hands off the ship they have been able to reinstate the mast and antenna, re-rigging it before erecting and stabilising it with robust cabling and a line that runs into the water for earthing it on the seabed. Once all the others vacate the hut, there will be a great deal more space to set things up properly. Already Jeffryes has communicated an intention to relocate the indoor wireless equipment from the workshop into the main room of the hut, where it is hoped the slightly warmer ambient temperature will be more conducive to sending and receiving. For the men staying on, such confidence is a boon. Perhaps they will be able to send messages back and forth with family and receive news of the outside world during their enforced second year.

Captain Davis is restless. He must get away. The hut at Cape Denison has been resupplied. He does not have to worry about the men who will stay on; they have more than enough of everything for a second year on the ice. On the other hand, Frank Wild and his men are in a precarious situation. Having made their base on an ice shelf some 1900 kilometres along the coast to the west, Wild and his men don't have the wherewithal to survive another year. Captain Davis must get to them or they will likely perish.

His search along the shoreline for Mawson's party proved fruitless. Skirting the ice cliffs in the *Aurora*, the captain and crew scanned the vast blank face of Adélie Land, desperate for any sign of the missing men, yet they saw nothing that would suggest any human presence at all. After three days, the ship returned to Cape Denison. Already Captain Davis has waited with the vessel for two weeks. He can wait no more. The season is getting on.

However, the weather is fast deteriorating and conditions are highly unsuited for the loading of men bound for home. No one in his right mind wants to brave hurricane-strength winds and heaving seas in the little motor launch. Fierce gusts from the south push the *Aurora* out to sea, which is probably for the best. Finding a safe anchorage in the Commonwealth Bay is not so easily done in foul weather and poor visibility; and with the ship struggling so, its steering often unmanageable, the hidden reefs and many islets closer to shore pose a diabolical hazard. With all the manoeuvring, the ship's coal supplies are dwindling. It is another concern to add to the captain's list of woes.

The tempest lasts for days and looks set to continue for an entire week. Fearful of what trials the captain and crew are facing out on the water, the men venture from the hut every few hours and conduct an anxious surveillance of the horizon. Occasionally from the swirly opacity of seaspray and driving snow there's a glimpse of something ghostlike, an ice ship, white as the landscape. Meanwhile on the bridge, a careworn and weary Captain Davis battles on, refusing to give in to the storm.

There is nothing exceptional about the circumstances. This is Cape Denison showing its true colours. It is not anything the men on dry land have not witnessed a hundred times before. Yet their hearts are in their mouths. Not only do they fear the worst for Captain Davis, they are all too aware that whatever vile weather they are facing on the coast, Mawson, Mertz and Ninnis are sure to be experiencing an even greater level of unpleasantness. At least in being confined to the hut they are enjoying a wonderful array of fresh foods. They're eating like kings.

CHAPTER FORTY-THREE

Mawson's lips are riven. The cracks are deep and painful and bleed when he eats. He doubts he could smile, even if he had a good reason. Windburn sores and the blister crusts from repeated frostbites bloom on his face and refuse to heal. His eyes are dry and bloodshot but he is pretty sure that he can see Commonwealth Bay in the haze to the north – it could be the sea stretching out under towering cumulus clouds.

Repeatedly delayed by poor visibility, he has come to the conclusion that he should probably have made his way to the sea and followed the coastline. It is all a matter of conjecture. Had he followed another course, he might have encountered other issues. Placing one foot in front of the other is all he can do in any direction. His steps are tentative on the slippery surface. And the strong side wind is doing what it can to unbalance him. The repeated falls are painful; his emaciated body is bruised black and blue. Every time he hits the ice he half expects to find a bone jutting through the skin.

He has less than a kilo of food left. He had calculated that it would be sufficient to get him to Aladdin's Cave, which he estimates to still be more than 30 kilometres away. Despite upping the distances he has been able to cover, he doesn't dare hope that he will reach the cache now. Recently he has been thinking about building a cairn, and placing his diaries and logbook in it with his belongings. Then at least everything might be found. That might have to be good enough.

In the event, it isn't necessary. From the gloom, a dark shape catches his peripheral vision. Had he not lifted his gaze momentarily from his feet, he would have missed it. Excitedly, he strikes off. It hardly needs to be investigated, this thing that from several hundred metres away is so recognisably man-made. Over recent months he has grown accustomed to distortions, tricks of the light and of perspective that white monotony is apt to play on a tired mind – he has learned not to trust what he sees. But getting closer, it is unmistakeable. It's a cairn, a pile of blocks of snow, its square shoulders wrapped in black fabric. The sight is quite jarring; he has grown so accustomed to travelling in a light that casts no shadow. This is a concrete thing. This is hope. This is salvation.

He seizes on the food left by McLean, Hurley and Alfie like a wild animal, tearing into the ration bag to see what delights it contains. Pemmican! Chocolate! Milk powder and tea and blessed, fatty, creamy, delicious butter! Words can't describe his joy at such a discovery. His hands shake as he unfolds the note that they have left, and the news contained within it is almost as greedily consumed. He reads the lines of the message, rejoicing at each word. Four words in particular: The ship has arrived. It hurts like crazy but it is impossible not to smile. What fine men!

News of Amundsen reaching the South Pole has the effect of buoying him. Perhaps he should mourn the loss of the prize that Scott fought so valiantly to claim for Great Britain. The trouble is, Mawson's spirits have been so lifted by the cairn and its treasures that he is inclined to celebrate all things and all people, regardless of nationality.

Most crucially, the note confirms his position. And confirms that he is on the right course. It is particularly useful to have directions to Aladdin's Cave – a distance of 33 kilometres. It's doable with the extra food. But he needs some good luck too. As tempting as it is to put up the tent and to cook up a meal and eat his fill, Mawson decides to make an immediate start. Besides, he knows that it is not a good idea to overload one's stomach after starvation.

McLean's directions are a little ambiguous. Mawson cannot be sure they are the right bearings. McLean states that the cairn lies S60°N from Aladdin's Cave. Presumably that means that from the cairn he should proceed N30°W. Or should he proceed along N60°W? He strikes out first along N30°W, then corrects to N60°W.

Crevasses are an ever-present threat, but his greatest concern is the slippery surface. It will make the going tough, particularly if there is any wind. If only he hadn't thrown away his crampons, nothing would stop him galloping to the finish line. But it will take him two days at least to swallow the distance. The thought of being able to rest and recuperate for a few days without having to worry about food is bliss. He prays for a calm day.

CHAPTER FORTY-FOUR

His new crampons look a little strange. Blocky and coarse looking, they have taken the best part of the day to make. Mawson cannot help but feel proud of his handiwork. That he has managed to fashion anything at all using his few simple tools and whatever meagre materials he has to hand is an achievement worth celebrating.

He takes a moment to admire his creation, made from sections of the wooden theodolite box, candlewick strapping and all the nails and screws he could find – some from the sledge itself, some from the cooker box, some borrowed from the sledge-meter. Strapping them on, he discovers that his hand-crafted crampons are not entirely effective. But in the absence of anything better, they will have to do.

Limping along in his new footwear is a trial. He is steadier overall, but the crampons are unbalanced. The right one is not working as well as the left. Consequently he is favouring his left leg – the muscles tense to hold himself upright as he swings his right around. Each time he tenses his body, he anticipates a fall. Taking the weight, his right leg splays out at an awkward angle as the chunky angular crampons slide across the uneven surface. It is a jarring and exhausting way to proceed.

By midnight, both crampons are broken and he finds himself on a slippery slope. Realising he cannot go any further, Mawson locates the only patch of snow on the icefield and pitches his tent. He feels rather broken too. His right leg is strained. But as the haze lifts, the ocean comes into view and there are the Mackellar Islands. And, best of all, at a distance of around 5 kilometres – another cairn. He knows from experience that distances in Antarctica can be deceptive, but this doesn't marry up with McLean's directions. By his reckoning it should be further away. He scarcely dares to hope. He supposes that he should rejoice. But he wants to cry. If this is Aladdin's Cave, how on earth will he get there across this ghastly glassy surface.

Repairing the crampons keeps him in the tent the whole of the next day. They are still woefully inadequate, but they don't need to carry him far. He feels certain that there will be spare crampons in Aladdin's Cave. He has overcome such obstacles; providence will undoubtedly see him through the next stage. Aladdin's Cave is tantalisingly close to Cape Denison. And he knows the way. He could walk it blindfolded. How many times has he hiked up and down that stretch of the ice ramp – his gaze forever directed outwards to the plateau! Now he is on the homeward stretch, his eyes will be fixed northward. With a good pair of sturdy crampons on his feet, the walk down to the hut will be a pleasure. And to see the hut, the men and the ship! With a clear mind he sets off.

CHAPTER FORTY-FIVE

A pineapple! And oranges! Is he dreaming? Bursting through the entrance to the snow cave, Mawson drinks in the sight of fresh fruit. The colours are so vivid, they appear to swim before his eyes like wild hallucinations. His heart is full of the greatest joy and thanksgiving. Scrambling in, sliding about on hands and knees, he can scarcely believe he has made it. He flops onto his back and breathes a deep sigh of relief.

Good weather and a light breeze has hastened his route across the last few kilometres, his own excited breaths echoing loudly within his head. It has been a tortured crossing from last night's camp, with lots of skidding and falling onto his backside; once again his crampons have utterly disintegrated – they have lost their straps and sit wildly askew, held fast around the ankle of his finneskoes. He couldn't care less.

True to its name, Aladdin's Cave is full of the most unbelievable treasure. Food in abundance. Fuel for the stove. The fruit fresh from the ship. The walls of his wonderful refuge glitter and sparkle in the light from the stove as he slowly defrosts an orange, savouring the sweet smell of its skin as it warms up in the glow. It is an impossibly beautiful thing in his frostbite-grizzled hands, the hideously damaged fingers with their weeping cuts and wounds and suppurating sores.

He stretches his sleeping bag out on the floor; a more welcoming and accommodating space he could not imagine. Except the hut of course! But that is only a few hours' march away. Never again will he need to struggle against an unforgiving wind to pitch that shelter, that ridiculous excuse for a tent. *One more night and I will be home*, he thinks as he drifts off to sleep.

However, the next morning his hopes are dashed. The blasted winds of Cape Denison make an icy descent to the hut utterly impossible. He would be blown off his feet immediately. Unfortunately there are no spare crampons. He will have to make another pair.

CHAPTER FORTY-SIX

Oh, for a clear spell. It has been four days of frustration. Every night he goes to sleep, full of hope that the next day will offer favourable conditions. That he will get away with his pair of freshly cobbled-together crampons strapped to his boots with a spare dog harness. Every morning after breakfast he gets himself ready to go as the wind rages outside, whistling and howling beyond the opening to the snow cave. And he sits.

He waits a day. He waits a night. He gets up at 3 a.m., checks again at 9 a.m., then decides at 11 a.m. that another day must pass him by. He is all too aware that the Cape Denison bad weather can last for a week or more. Will the ship wait for him? The question burns. If the *Aurora* has left, will there be anyone at the hut? Perhaps he has been given up for dead. Is this to be his fate? To remain alone, a prisoner in this cage of white fright and fury?

Arriving at Aladdin's Cave, Mawson felt transported with feelings of exuberance and the purest joy; but now that he is stuck here, to leave it would represent an even greater happiness.

He prays for a few hours of calm, a lull, a clearing. No chance. The spiralling, gusting, whirling white chaos persists. It throws the sledge end over end, then buries it completely. Good riddance, Mawson thinks sourly.

At least there is food. Slowly he has increased the amount of food he is eating at meals. He practises eating. He pauses to make sure it agrees with him. Dunking sledging biscuits in hot milk mixed up from Glaxo powder has become a favourite pastime when there is nothing to do but eat. Gaining confidence in his appetite, he levers great chunks of pemmican out of the tin. He chews slowly, not savouring the richness. He wears a look of boredom. His determination is being slowly replaced by despondency.

These biscuits are horribly stale, he thinks gloomily.

Every time he takes a bite he thinks of hunger. He is no longer hungry, but the hunger has stayed with him. A week ago, such a mouthful would represent an entire meal. He would not have cared if it were stale or crawling with weevils.

Now he is beginning to wonder if this food isn't spoiled. He has a sneaking suspicion that his diet is slowly poisoning him.

In a troubling new development, his nose appears to be haemorrhaging. Thin, watery blood dribbles in a slow stream from one nostril. He cannot staunch the flow. It is as though his inner resolve is disintegrating. To make matters worse, his limbs feel suddenly very heavy and his joints ache to an extent that keeps him awake at night. He hopes it isn't the early signs of scurvy.

CHAPTER FORTY-SEVEN

They stand on the rocky ridge to watch the *Aurora* weigh anchor – Madigan, Dr McLean, Bage, Hodgeman, the new man Jeffryes, and Bick, who has attached a Union Jack to a bamboo pole. He swings it back and forth in reply to the blue ensign that the crew on the ship have dipped on the ship's rigging in farewell. It's a melancholy gesture. Quiet conversation passes between them, mostly concerning how lucky the men heading for home are. There is also some discussion about the weather and how it might affect their soon-to-be-relaunched search efforts. Is hope fading? The low tone of their voices betrays a certain tentativeness. How credible are their chances of finding the overdue men?

And what challenges lie ahead for them; how will they themselves fare?

They stay for a while, watching, the ship getting smaller and smaller as it heads for the vanishing point on the northern horizon. It is early afternoon and the sun floods the landscape in bright white light, although the nights are getting much darker. For months they have dreamed of leaving here, counting down the days until the ship's return in January. Now it is 8 February, 1913, and the calendar countdown will have to start all over again. They won't be home until 1914.

'See you in ten months,' says Alfie rather wistfully to the disappearing ship. Soon it will be a tiny speck.

Madigan sighs. 'Only three hundred and four days to get through.'

'What's for lunch, then?' asks Jeffryes.

Jeffryes, the new wireless operator, is far from forlorn. He is excited to be spending a year in Antarctica, where everything is new and exciting and strange. It's a dream he never thought he would fulfil, not after being turned down by Mawson in favour of Hannam, way back when the expedition was still in its planning stage.

The others listen to Jeffryes's enthusiasm and are reminded of their own delight when they first stepped foot in Cape Denison. Now the landscape reminds them of their frustration and resentment. They fear that their search for

Mawson's party will prove fruitless, and how will they get through another winter in this miserable place?

They do feel a certain amount of pride too – one must do one's duty. After all, nobody has forced them to stay on at Cape Denison. They are volunteers – some more willing than others. Madigan confides to his diary that he would have felt a coward and a deserter for the rest of his life had he refused to stay as Captain Davis had asked. It's a good thing that the six men get along well, but Madigan wonders what he will find interesting enough to note down in his diary. Hut life is going to be rather more sedate without some of the more interesting and infuriating characters gone.

It has been a busy few days packing away all the new stores, stockpiling seal meat and penguin eggs in the verandah, stacking fuel and lining the exterior of the hut with canvas to make it more weathertight. Extra wire lines have been lashed across the roof and secured to the rocks around the hut. Nobody wants to take any chances. They've survived one winter onslaught, but there are no guarantees the hut will withstand another one.

Jeffryes has made some improvements to the wireless set-up. He has covered the rocks around the radio mast with wire mesh, which he says will strengthen the signal. He has discussed running steel cables down into the Boat Harbour to earth the wireless. They will complete the work today if the weather holds. Already they can tell that Jeffryes has the skills to get everything sorted. If they can send and receive messages, then at least they will not feel so cut off from the world.

With so few of them living in the hut, there's not the same need to maintain a strict roster for routine tasks or chores. It will largely be up to each man to decide how to occupy his time. Madigan will continue his meteorological work, and Bage will carry on with his magnetic observations. Whoever feels like cooking will be chef for the day. There won't be a clamour of interest for any sweeping, cleaning and washing up but they'll all take turns depending on who is around and who is available. One exceptionally good bit of news is, with so few men in the hut, they will get to bathe whenever they like. Whether they will or not remains to be seen. Washing every eighteen days has become somewhat of a habit. Nobody has seen Jeffryes take a bath.

Madigan has elected to look after the twenty-one dogs that Amundsen sent down with the ship from Hobart, at least until Mertz and Ninnis are back from sledging. It was a generous gift, yet there are simply too many to look after. Already there are three puppies, pushing their number up to twenty-four. Madigan and Bick have discussed the difficulty of caring for them over the brutally cold winter. It is not fair to subject them to such hardship. Unfortunately, a grim fate awaits at least half of them.

After lunch, Madigan, Bick and Jeffryes return to the harbour to place the wireless cables. They haven't been working long when Madigan shouts. 'Hey, look! Look! Look!'

Bickerton straightens up and stares. A figure hauling a sledge is slowly making his way down the distant slopes from the plateau. 'My God,' he says. 'Who is it?'

'Who is it?' Madigan echoes Bick's question. 'It's Mertz.' But he's not sure. He wishes he had binoculars.

'One man. Where are the other two?' Despite not knowing the men in question, Jeffryes, too, is startled by the sight of a person appearing from nowhere.

Madigan is excited. 'The others must be further back. Maybe hauling the other sledge.'

'But the dogs,' says Bick. 'Where are the dogs?'

'Not tall enough to be Ninnis.'

Bick starts to wave. 'No, it's Douglas. It's definitely Douglas.'

'Yes, it's Mawson!' shouts Madigan as he starts to run towards their leader. 'Recall the ship, Jeffryes!'

Bick is running too. 'We might be going home after all!'

Jeffryes stays rooted to the spot, experiencing a tinge of disappointment at the thought of going home so soon. Yet he cannot help but feel moved at the sight of this lone man, with both arms raised over his head in a slow, silent greeting.

EPILOGUE

Mawson returned. Ninnis and Mertz dead.
Return immediately and pick up all hands.

This was the simple message conveyed to Captain Davis aboard the *Aurora*. Would they receive it in time? The ship was only five hours from Cape Denison when Mawson made his miraculous return. Before leaving, Hannam had agreed with Jeffryes to listen for news. Every night from 8 p.m. to 1 a.m. he was to tune in to the receiving set onboard. Hannam did not have the apparatus to send messages back, unfortunately; therefore he could neither confirm that they had intercepted the message, nor report that the ship was coming back to retrieve them.

Several anxious hours passed while the shore party waited. Would the *Aurora* reappear? In the meantime, Mawson recounted the shocking tragedy of Ninnis's disappearance down the crevasse, and of Mertz's decline and eventual death after struggling homeward with very little food and no proper shelter. Mawson admitted that at times he had held out little hope of surviving himself. The story left the men reeling with emotion. For Madigan and Bickerton it was particularly heartbreaking to think that they would never again see their great friends from Hyde Park Corner.

When the ship reappeared in Commonwealth Bay first thing the next morning, it was a huge relief. It seemed that deliverance from their miserable enforced stay was imminent. The men had spent much of the night packing up their personal effects and essentials. They were more than ready to depart. Yet the relatively calm conditions that had persisted throughout the day of his arrival and into the evening were not to last. Right on cue, the Cape Denison weather underwent the all-too-familiar deterioration. Ferocious winds kept them confined to the hut even as the *Aurora* steamed to and fro, hoping for even the briefest break in the wild storm so that they may launch a whaleboat.

Another message was sent to the ship:

Anxious to get off, hope Captain Davis can wait a few days longer.

But Captain Davis could not wait. With winds so strong, the ship's engines were at maximum power just ensuring the drift did not push them onto the reef. As reluctant as he was to contravene orders from Mawson, Captain Davis realised the heavy weather could last a week or more. The safety of the ship and its crew, and the safety of Frank Wild's men to the west were at stake. To stay on any longer would be courting disaster. Given the shore party had everything in place for another year, he felt justified in prioritising the needs of the others, who were in far greater need of rescue with autumn fast approaching. Already the *Aurora* was three weeks overdue.

Douglas Mawson was in a severely debilitated state. His legs were horribly swollen, he was covered in sores and raw patches where skin had been rubbed off. His strength was all but gone, and his digestion badly affected by weeks of poor diet. Even in late March his condition remained shaky. In a diary entry he says, *'my nerves in a very serious state, and from the feeling I have in the base of my head I [have a] suspicion that I may go off my rocker very soon. My nerves have evidently had a very great shock.'*

Mawson later suggested that he may not have survived a long, rough sea crossing aboard the *Aurora*. Forced to stay in Cape Denison, he was able to rest and recuperate under the expert medical supervision of Dr McLean. In the months that followed he did not often talk about his harrowing ordeal. Neither did he dwell on the deaths of his two companions. Madigan in particular found this to be a rather callous attitude, yet these memories were evidently too painful for Mawson to discuss, and he preferred to focus on the task of writing up his scientific work. Another source of stress for Mawson was undoubtedly the cost associated with a second year in Antarctica. He would be returning to Australia with significant debts to clear and a book to write.

Much to everyone's relief, two-way wireless communication was soon established between Cape Denison and the relay station on Macquarie Island, and regular messages were sent on to Sydney, Melbourne, Wellington and Hobart. Condolences were sent to the families of Ninnis and Mertz, as well as to Kathleen Scott, the widow of Captain Scott who, they had subsequently found out, had perished on his return journey from the South Pole.

Where Hannam had failed, Jeffryes had triumphed. However, the new wireless operator was soon to become a liability rather than an asset. Psychologically unstable, Jeffryes did not cope at all well with being cooped up in the hut over the Antarctic winter. Both Madigan and Mawson were subjected to violent assaults when Jeffryes's growing levels of paranoia reached a tipping point. His increasingly erratic behaviour included refusing to wash and storing jars of his urine on the shelves beside his bed. At one stage, he even accused Bickerton of loading a gun to shoot him. The dangers of living with someone whose mental state was so volatile were obvious, yet there was nothing they could do. They certainly could not turf him out of the hut. Even Dr McLean was at a loss.

All the while, Jeffryes remained their sole radio operator, responsible for sending and receiving messages. Obviously, as the only link to the outside world, someone experiencing such a catastrophic mental collapse could not be trusted. Indeed, Mawson discovered that he had started to send out rogue messages, including one in Mawson's name declaring that five of the seven men had gone insane. During one of the man's moments of clarity, Mawson was able to convince him that his workload was taxing him unnecessarily. Jeffryes agreed that Bickerton should take over his wireless duties. His erratic behaviour continued into the springtime, but by summer he seemed to be coming right.

During the final weeks of their stay, Mawson, Madigan and Hodgeman set out to see if they could retrieve some of the expensive scientific equipment that had been jettisoned on the previous year's journeys by the eastern, western and southern sledging parties. It had all been on loan from the Admiralty and the Royal Geographical Society in London. They came back empty-handed. The Antarctic snow drifts had buried everything – their sledging tracks, their caches and all trace of their friends Mertz and Ninnis.

The *Aurora* returned to pick up the men on 13 December. Nobody was sad to leave Cape Denison, the windiest place on earth, where the average annual windspeed turned out to be 80 kilometres an hour. Some days the wind exceeded 145 kilometres an hour, with gusts from the interior reaching a staggering 320 kilometres per hour.

Their sea journey back to Australia was an arduous one that included a significant detour to carry out oceanographic work to the west. A more direct route would no doubt have been preferred. When they finally made landfall in February 1914, the men's relief must have been enormous.

The Australian Antarctic Expedition's achievements were many – both in terms of scientific results covering meteorology, geology, biology, bacteriology, and the earth's magnetic field; and in the explorations of hitherto unknown regions. Their combined sledging journeys covered in excess of 6400 kilometres – most of it achieved in appalling weather. Yet it is a sad fact that much of the important work that was completed during their time in Cape Denison was overshadowed by the news of Robert Falcon Scott's tragic death.

Douglas Mawson's tale is one of the most legendary in polar history, an epic struggle between one man's self-belief and the worst conditions the hostile polar environment can throw at him. His solo journey represents not only a feat of physical endurance but also a triumph of the human spirit. With the help of Dr McLean, Douglas Mawson wrote about his death-defying experience in *Home of the Blizzard*, which was a publishing success. The funds the book generated were of enormous assistance in offsetting the expedition's sizeable debts.

Perhaps against his better judgement, Mawson returned to Antarctica in 1929. Thankfully, the British Australian & New Zealand Antarctic Research Expedition (BANZARE) carried out its significant program of scientific work without serious incident.

Two hardy souls proved even keener to get back on the ice than Mawson. On returning from Mawson's AAE, Frank Hurley and Frank Wild signed on to Ernest Shackleton's *Endurance* expedition in 1914 to cross the Antarctic interior from one frozen coast to the other. As fate would have it, neither man would set foot on the continent after the ship was caught in the sea ice and sank, leaving all twenty-eight men stranded on slowly deteriorating pack ice in the middle of the Weddell Sea. Luckily, both would survive to tell the incredible tale.

The two men who would never leave Antarctica, Xavier Mertz and Belgrave Ninnis, paid the ultimate price in the quest to bring form and meaning to an uncharted region of the continent. Their aim was to put this place on paper, so it is wholly appropriate that their sacrifice should be acknowledged on the map. Two landscape features will forever bear their names: the

Ninnis Glacier at 68°22′S 147°0′E and the Mertz Glacier at 67°30′S 144°45′E. It is a fitting tribute to two men who willingly put themselves in danger in order to extend geographical knowledge.

ABOUT THE AUTHOR

Joanna Grochowicz is a polar historian and author. Her previous narrative non-fiction books on early Antarctic exploration include *Into the White: Scott's Antarctic Odyssey*, *Amundsen's Way* and *Shackleton's Endurance*. Fast-paced and exquisitely written, they reveal in fascinating detail the human aspirations and tragedies that have shaped our understanding of the polar regions. Drawing on diaries, letters and expedition narratives from archives and personal collections around the world, Grochowicz brings to light new and often overlooked elements of heroic age exploration. Realistic portrayals of historical figures and the interplay of personalities lend her work a grittiness and authenticity that is not often present in historical writing for children and young adults.